550
C593f

San Diego Christian College
Library
Santee, CA

CHRISTIAN
HERITAGE
COLLEGE
LIBRARY

PRESENTED BY
Mr. & Mrs. Jack T. Baker
IN MEMORY OF
Marvel F. Barnes

FOSSILS, FLOOD, AND FIRE

OTHER BOOKS BY HAROLD W. CLARK

Back to Creation, 1929

Genes and Genesis, 1940

The New Diluvialism, 1946

Creation Speaks, 1947

Nature Nuggets, 1955

Skylines and Detours, 1959

Wonders of Creation, 1964
Pacific Press Pub. Assn., Mountain View, Calif.

Evolution and the Bible (Filmstrips with scripts), 1965
Review & Herald Pub. Assn., Washington, D.C.

Crusader for Creation, 1966
Pacific Press Pub. Assn., Mountain View, Calif.

Genesis and Science, 1967
Southern Pub. Assn., Nashville, Tenn.

Note: The first six are out of print.

Fossils, Flood, and Fire

by

HAROLD W. CLARK

Professor Emeritus of Biology
Pacific Union College
Angwin, California

Published by OUTDOOR PICTURES, Escondido, California

COPYRIGHT, 1968, BY
HAROLD W. CLARK

Library of Congress Catalog Card Number 67-29223

Printed and bound in the U.S.A.

Contents

Chapter		Page
	Foreword by Ernest S. Booth	6
	Preface	7
	Part One—Historical Survey	9
1	The Authority of Science	11
2	The Rise of Modern Geology	21
3	The Delusion of Uniformity	29
4	The Flood Theory of Geology	37
5	Flood Legends	45
6	Ancient Life Zones	51
	Part Two—The Rocks and the Flood	61
7	The Basement Complex	65
8	Life of the Ancient Seas	77
9	Burial of the Coal Forests	91
10	Permian Puzzles	105
11	Mystery of the Red Beds	111
12	The Violent Climax	127
13	Shaping the Landscape	143
14	The Reign of Winter	169
15	Cave Men and Stone Ages	205
16	This Changing Planet	227
	Index	234

Foreword

For more than a century geologists have worked to explain the strata of the earth in terms of long ages, by processes which have acted over a period of more than 500,000,000 years. Today virtually every geological formation on earth can be explained in detail by one or more hypotheses based on uniform conditions throughout geologic time. One might feel, then, that another book on the subject is hardly needed at this time.

Yet, no matter how detailed were the explanations for geological phenomena, none had attempted to go into detail regarding the Universal Flood described in Genesis, and found in legends of nearly all primitive peoples, as the cause of most geological activity in the past until the publication of *The New Diluvialism* by Harold W. Clark in 1946. Since only a few copies of this book were printed, it never experienced a wide circulation.

Now, 21 years later, Professor Clark has written a new book, based on the former publication, to show how most geological formations can be interpreted in terms of the Universal Flood. Without question, FOSSILS, FLOOD, AND FIRE is the most complete account of the effects of the great cataclysm ever written.

Professor Clark, now retired, has spent a lifetime studying geology. His course in geology was always a favorite with students of Pacific Union College in California, where he taught for 37 years. It is my conviction that this new book is a major contribution to the literature of geology.

ERNEST S. BOOTH

Preface

For many years men who accept the Genesis record of the Flood have written in opposition to the popular theory of long ages of geological time. No one, however, to my knowledge, has made any attempt to portray the geological actions of the Flood in systematic order. George McCready Price wrote the *New Geology*, but it was purely descriptive as far as geological data were concerned. But in this study we intend to make an attempt to show how the sequence of events during the Flood can be understood in the light of geological data.

Doubtless the main reason why a more positive interpretation has not been made in the past is that sufficient geological data has not been available upon which to build a positive scientific study in line with the Genesis record. Perhaps another reason has been the fear that such an effort would be regarded as indulging in mere speculation.

It seems to me, on the other hand, that the time has come for someone to make an effort to align with the Flood theory of geology rather than with "ages geology," the vast amount of geological knowledge that has accumulated in recent years. And so, in the following pages we are taking the risk of being branded as unscientific, a wild speculator, or a dreamer, in order to give a new viewpoint to geological interpretation.

The reader should understand clearly that the author does not make any claim to infallibility. Interpretations which are suggested may be open to challenge. But, if out of this effort can come a recognition of the fact that geology can be interpreted in terms of the "short chronology" of Genesis just as well as, or even better than, the "long chronology," then the publication of this treatise will not have been in vain.

Fossils, Flood, and Fire

In preparation of this material the author has read tens of thousands of pages of geological literature, and has tried to separate the facts from the assumptions, or from biased conclusions that arise from a century and a half of the teaching of uniformitarianism. He has also traveled widely, in every state of the Union and most of the Canadian provinces, searching for evidence. He has weighed carefully and judiciously the opinions of his diluvialist colleagues, and has attempted to construct a conservative, consistent interpretation of geology that might eventually lead to a real science of "Flood Geology" or Diluvialism.

The reader may be tempted to ask, when he reads some of the statements regarding action in different stages of the Flood: How do you know that it really did happen that way? In answer it must be pointed out that we are merely stating in a positive manner the best interpretation we have been able to offer. If anyone has a better one, it certainly is welcome. This study is simply a beginning along the line of positive approach to the problems of geology and Genesis.

Doubtless some will be disappointed that there is no documentation for the material given in the following treatise. But a list of references has been omitted in the interest of brevity and readability. In preparation of this book literally hundreds of books and journal articles have been consulted. To refer to all sources from which data have been drawn would make the reading bulky. Probably not many readers would care to wade through a mass of technical material, anyway. And so, in order to make it easier for the average person to follow the line of argument, this has been omitted.

HAROLD W. CLARK

Angwin, California

PART ONE

Historical Survey

Before we take up a discussion of the problems associated with geology and the Flood, it is important that we understand the historical background. Accordingly, in the first three chapters we shall discuss the events in the history of geology, and related sciences when necessary, that have led up to the present status of the science of geology.

Some ideas that have shaped the science of geology are very old, particularly those in the field of philosophy. Therefore we must go back to antiquity in order to trace the development of modern concepts. Then, too, we must challenge the validity of scientific theories, as there may be a question as to whether they are always dependable. Furthermore, there is that perplexing question of the relation between science and revelation. After these problems have been discussed, we shall be in a position to outline the problems of geology in the light of *Diluvialism,* or "Flood Geology."

CHAPTER ONE

The Authority of Science

A visitor to any museum of natural history stands awe-struck at the display of mounted skeletons of enormous animals which lived in prehistoric times. He can but marvel at the virility once possessed by nature when dinosaurs more than eighty feet long roamed the earth. How and when these great creatures lived and how they were exterminated are questions that almost universally arise in the minds of thoughtful students.

To some the past may seem inconsequential; but careful thought will reveal the fallacy of this attitude, for the future is bound up largely with the past. It was formerly accepted as sufficient to say that the Genesis story of Creation and the Flood gave us an explanation of these things, but of late years there has been a growing effort to explain everything, past and present, by natural causes. Evolution has been accepted, not because it has been proved, but because as a theory it seems to give a general scheme by which the past can be explained. On the other hand, creationism has been largely relegated to the limbo of discarded theories, and considered unworthy of the attention of intelligent

11

Fossils, Flood, and Fire

men and women. And along with this rejection of the Genesis story of creation has gone its corollary, the account of the Flood.

In spite of this, thousands still accept the Genesis record of Creation and the Flood as an inspired revelation from God to man. And in accord with this record, they understand the earth to have been in existence only a comparatively short time instead of having passed through long geological ages.

The truth of the matter is that creationism is one of the oldest of all recorded explanations of the origin of the earth and its life. The book of Genesis was written a thousand years before the Ionian philosophers formulated their naturalistic cosmogonies. For over three thousand years it has been regarded as an authoritative statement regarding the beginning of the earth.

For many centuries the creation doctrine was the orthodox belief of the Hebrew people, and served to distinguish them from the pagan nations around them. The Exodus from Egypt separated them from the idolatrous worship of nature gods, and brought them back to the worship of the true God, the Creator of the heavens and the earth. Throughout 1500 years of national history this belief in God as the Creator was kept intact, even though at times false philosophy threatened to destroy it.

The gods of the ancient nations, with the exception of Jehovah, the God of Israel, stood as personifications or symbols of natural forces, which were self-operating and independent of any divine or supernatural intervention. Matter was assumed to be eternal, and the heavens and the earth, even the gods themselves, were supposed to have arisen from the continuous and orderly processes of nature. Any such idea as a supernatural creation seemed to be foreign to their thinking, as the following quotation shows:

The Authority of Science

Probably the idea of creation never entered the human mind apart from Revelation.—Catholic Encyclopedia, Art.: *Creation*.

While in some of the ancient religions there appears to have been some lingering remains of the recognition of one God, we get an idea of how this situation came about in the statement made by Paul in Romans 1:21-23:

> Because that, when they knew God, they glorified him not as God, neither were thankful, but became vain in their imaginations, and their foolish heart was darkened. Professing themselves to be wise, they became fools.

The nature worship of ancient pagan nations may therefore be considered as a degeneration of monotheism, with the transfer of the attributes of God to nature itself. Accordingly, the original creationist view associated with faith in one Supreme God, became, by transfer to natural forces, the philosophy of evolution, since nature was assumed to be possessed with inherent forces and tendencies that drive it onward to perfection.

During the sixth century before Christ the intellectual leadership of the world passed from the Euphrates Valley to the shores of the Aegean Sea. With characteristic Greek logic the Ionian philosophers remolded the learning of Assyria and Egypt and developed a thorough-going naturalistic philosophy. All faith in the gods was challenged, and men were left in a skeptical frame of mind. Then came the great systematic philosophers, Socrates, Plato, and Aristotle being the greatest. These men and their collective thinking led others to try to solve the questions of the nature of God, His relation to the world, and the problems of politics, religion, and morals. Plato and Aristotle taught theories that later were to play an important part in the development of Christian theology. Plato said that God

13

Fossils, Flood, and Fire

was supreme, but beyond this one point his ideas regarding God were so vague and his interpretation of nature so pantheistic that his philosophy was of little value. Aristotle, on the other hand, leaned toward a more pragmatic attitude, and thought that he saw in nature all that was needed to explain its complex relationships. According to his philosophy, the whole creation moves toward a final ideal of perfection. Natural processes are controlled by an inner "necessity" or driving force, toward the fulfillment of an ultimate destiny. This is due to the presence in nature of inherent properties rather than to the direction of any higher power.

Greek culture as it spread over the world after the conquests of Alexander, came into contact with Judaism. Many of its philosophical concepts were accepted, with the result that Greek teachings became incorporated into the dogmas of some of the Jewish sects. Greek influence from the Alexandrian school in Egypt became a powerful factor in spreading Greek culture in Palestine. The knowledge of God was so effectually shut out by the imaginations of men that the true God was almost lost sight of, and many of the Jews worshipped a god of Greek origin.

Christ did not attend the rabbinical schools with their liberal teaching which had developed as the result of Greek influences. He learned His theology from the Old Testament writings; and in the work of His disciples may be seen the same truths that were taught by the holy men of old.

As long as the Christian churches retained the simplicity of the early gospel, the world witnessed a revival of morality and religious fervor beyond anything ever known. But when Christianity became popular, the spirit of critical inquiry and the study of philosophy led scholars to organize and codify the Christian teach-

The Authority of Science

ings. Origen and Clement were the outstanding leaders in the Eastern churches. Since they could not vindicate the Scriptures against the philosophy of the day, they introduced mystical interpretations of Scriptures like those that were being taught by the Platonists in Alexandria. In the Western churches Augustine followed a similar method. His views are of particular interest to us, as he attempted to explain creation in a way that would agree with both the Hebrew Scriptures and the Greek ideas. He recognized the God of the Bible as the supreme Creator, and taught that the world was created *ex nihilo,* that is, from nothing. At the same time he adopted the Greek view that creation was not complete, but that all nature was endowed with inherent power by which it might undergo development. Note this statement:

> 'We must be on our guard against giving interpretations which are hazardous or opposed to science, and so exposing the word of God to the ridicule of unbelievers.' (Quoted from Augustine.) An admirable application of this well-ordered liberty appears in his thesis on the simultaneous creation of the universe, and the gradual development of the world under the action of the natural forces which were placed in it. Certainly the instantaneous act of the Creator did not produce an organized universe as we see it now. But, in the beginning, God created all the elements of the world in a confused and Nebulous mass . . . and in this mass were the mysterious germs . . . of the future beings which were to develop themselves, when favorable circumstances should permit.—Catholic Encyclopedia, Art.: *Augustine.*

With the fall of Rome, the intellectual leadership again shifted, this time to the southern shores of the Mediterranean, from which it spread out with the later development of the Saracen crescent, into a broad belt reaching from Baghdad to Spain. During the 8th to 12th centuries Arabic schools were organized on the pattern of the Greek, and they published commen-

Fossils, Flood, and Fire

taries, encyclopedias, dictionaries, and scientific manuals. Plato, Aristotle, and the Neo-Platonists from Alexandria inspired the cosmology and the fundamental scientific approach in these schools.

Probably the most illustrious scholar of the Arabic world was Averroes (1126-1198), who lived in Morocco, Seville, and Cordova. He held Aristotle in great esteem. He taught a dual doctrine, that religion and philosophy are separate. Religion is only symbolical, and is for the ignorant masses, whereas philosophy is for the intellectuals. His philosophy was Aristotle over again — matter is eternal, and the world in an emanation, or extraction, not a creation.

Many European scholars attended the Arabic universities, and brought back their learning. Europe slowly absorbed the Arabic knowledge, and eventually there arose a demand for the establishment of universities in Europe. A considerable number of translations of Aristotle were made, and during the medieval period ancient classical knowledge from the Latin was welded with the Christian faith as interpreted in the light of Plato and Aristotle.

An important development came between 1200 and 1225 when the complete works of Aristotle were translated into Latin. Previous to this time Catholic philosophy had largely been founded on Platonic ideas, and now the church feared that the new ideas would undermine its authority. Then Thomas Aquinas stepped in and established an interpretation that has had far-reaching effects, not only in theology, but in science. He set up a dualism between science and religion, something like that of the Arabic scholars. Religion, he declared, belongs to the church, but men may believe anything they please in the field of science just as long as their ideas do not detract from their allegiance to the ecclesiastical powers. The only world we know,

The Authority of Science

said Aquinas, is the world we can learn by our senses. And so, what is to hinder us setting up a world of science based on observation, while we still render allegiance to the church in spiritual matters?

This notion was picked up later by Francis Bacon (1561-1626), who argued that religion must be based on the Bible, but that the Genesis record must not be used at all as the basis for any scientific conclusions. In this he was fully in accord with Augustine's teaching that all that Genesis implies is that God is the Creator, but how creation came about is to be left to science.

The Reformation surrounded the question of the relation of science and religion with a new atmosphere. This was the "back to the Bible" concept. As an essential part of the return to the literal rendering of the Bible came an interest in the creation record of Genesis. Although some of the reformers were perplexed by many of the confusing ideas that had been promoted regarding Genesis, yet the general trend was toward an acceptance of the record in its simplicity, both with respect to creation and to the Flood.

Not only were literal creationist views promoted among the Protestants, but the Catholic Counter Reformation under the leadership of the Spanish Jesuit Juarez established creationism as an orthodox doctrine of the Catholic Church, and this attitude continued until the swing toward evolution in the 19th century.

The period from the Reformation to the middle of the 19th century has been called the "Golden Age of Creationism." Many fundamental discoveries in science were made, and there was a genuine spirit of recognition of the validity of the Genesis story of creation and the Flood as a background for science.

However, as geological knowledge grew rapidly in the 18th century, theologians found it increasingly difficult to adjust the new knowledge to the short

Fossils, Flood, and Fire

chronology of Genesis. With increasing favor they began to turn to notions that were being propounded by scientists, not all of whom were sympathetic toward the Scriptural account of the past.

Buffon's theory of geological epochs in earth's history gave rise to the "day-age" theory—that the days of Genesis were only geological periods. Cuvier's theory of a series of catastrophes gave some theologians the idea that they could allow for the ages of geological time and the Genesis record at the same time. Then, when in 1840 Agassiz showed that most of the scattered debris in northern Europe and northeastern North America, which had been interpreted as remnants of the Flood, was actually glacial in origin, the Flood theory of geology soon died out.

It was revived half a century later by a number of writers, but of this modern revival we shall speak later. With it came a more complete study of the creation viewpoint, and in recent years there has been developed what might be termed the "New Creationism." This attempts to orient the origin of plants and animals with the Genesis record of creation in six literal days only a few thousand years ago, and the destruction of the world later by a universal flood.

Some object to this position, saying that the Bible is being placed on a scientific basis and being used as a textbook of science. To this objection it may be pointed out that the Bible is not a textbook, but that it does contain fundamental truths by which all scientific study must be oriented. If the Bible is to mean anything, it must be considered as inspired by the Spirit of God; and if inspired, it must be accepted as historically true. When, therefore, human speculative methods of thought assume to interpret the past history of the earth in a manner diametrically opposed to the Genesis record, Christians should take exception,

The Authority of Science

and take their stand on the Bible, as their guidebook.

The Bible must not be judged by men's ideas of science, for scientific theories come and go; on the other hand the Word of God abides forever, and human theories must be brought to the unerring standard of the Word. The Bible may not give details, they are left for man to discover; it does, however, lay down basic philosophical principles on which science is to be interpreted.

CHAPTER TWO

The Rise of Modern Geology

Up until the 17th century the people of Europe and America regarded the fossils as *lusus Naturae,* "Nature's little games." But in 1669 Niels Stenson wrote on the rocks of Tuscany and recognized the fossils as organic remains. Others followed his lead, and soon the idea was generally accepted.

The director of the Imperial Museum at St. Petersburg, J. G. Lehman, divided the rocks of the earth into three groups, (1) *Primitive,* or non-fossiliferous, (2) *Secondary,* or fossil-bearing, deposited during long periods of time in which plants and animals lived and died and left their remains, and (3) *Alluvial,* rocks which he supposed were left by Noah's Flood as it swept over the earth at the close of geologic time. This conception was taken up by other men of the time, elaborated, worked out more explicitly, and applied to the earth as a whole.

The French naturalist Buffon (1707-1788) is sometimes recognized as the most influential man of his times in developing geological theory. He combatted the idea of a universal Deluge, and attempted to describe the beginning, the past, and the future of the earth. His book *Epoques de la Nature,* published in 1778, outlined the seven epochs throughout the earth's history. Incidentally, this theory became the basis for the "day-age" theory adopted by many of the churches in the early 19th century.

Another influential writer was A. G. Werner, who for a long time was professor of mineralogy at Freiberg, Germany. He taught that all the rocks of the earth had originated as mechanical or chemical precipitates from waters that had enveloped the whole earth. Even volcanic products had been formed this way and later melted. His students came from all of Europe, and carried his ideas back home with them.

Werner distinguished five series of rocks that had enveloped the earth like the coats of an onion. Thus his theory has been dubbed the "onion-coat theory." Unfortunately his observations were limited to a small part of Germany and nearby areas, and he knew nothing of the rocks in other parts of the earth.

Geological theory in its modern form was introduced by James Hutton to the Royal Society of Edinburgh in 1785. Hutton had traveled widely and observed closely, and his conclusions had more scientific background than those of his predecessors. However, in attempting to explain the present state of our globe, he rejected the Flood concept and went back to the views of the ancient Greeks. After describing the globe, its core, water, land, and air, and the means by which the strata were supposed to have been formed, he declared that all these processes must have required a "time which is indefinite." He imagined one cycle after another, and concluded that

> The result, therefore, of our present inquiry is, that we find no vestige of a beginning—no prospect of an end.—*Theory of the Earth,* p. 204.

The Rise of Modern Geology

Hutton's presentation was so difficult to follow that little attention was paid to it until John Playfair published his *Illustrations of the Huttonian Theory of the Earth* in 1802. He said:

> The authority of the Sacred Book seems to be but little interested in what regards the mere antiquity of the earth itself.
> It is but reasonable, therefore, that we . . . (suppose) that the chronology of Moses relates only to the human race.

Thus at one stroke of the pen he swept away all scientific credence for the Flood, and opened the way for the acceptance of long ages of geologic time. The influence of his discussion was definitely to turn the attention of geologists in the direction of uniformitarianism.

About the time that Hutton was setting forth his views, a canal engineer, land agent, and surveyor, William Smith, made discoveries that were destined to play an important part in the development of geological theory. During his career he observed that certain layers of rock always occurred in certain relationships and always contained certain types of fossils. In 1815 he constructed a geological map of England. This map, which now hangs in the Royal Museum of Geology in Toronto, Canada, shows the different geological features in color. Many of the names assigned by Smith are still used. Because of his influence he has been called the "Father of British Geology."

Smith's findings came to the attention of the geologists, and as a result of the investigation of these phenomena there was soon developed a method of identifying rock strata and determining their supposed age by the contained fossils. Thus the principle of correlation by means of fossil contents became the basis of the science of stratigraphic geology and paleontology.

Fossils, Flood, and Fire

Previous to the time of Smith the only means of determining the relative "age" of the rocks was by means of their lithological character. Now, by *assuming* that great ages of time had elapsed, it was *supposed* that there had been a succession of life during those ages, and therefore the fossils could be used to classify the rocks and place them in their relative position continuously from one outcrop to another, and from one country to another; and since in many localities the rocks have been displaced from their original positions by folding, faulting, and other movements, so that it is almost impossible to trace their sequence, it was thought that by means of the fossils the rocks could be reduced to a system and their relative ages determined.

To Charles Lyell is generally ascribed the popularity of the uniformitarian theory of geology. His *Principles of Geology,* first published in 1830, ran through twelve editions and was used as a textbook for about fifty years. It brought together data from all over the earth, with the express purpose of showing that all past changes have been of the same nature as those now going on. He said:

> Amid all the revolutions of the globe, the economy of nature has been uniform.

The influence of this textbook, and his *Elements of Geology,* as well as other publications on geology, has been far-reaching. He traveled widely and made extensive observations, which he incorporated into new editions of his book as they appeared.

Castastrophism did not die out quietly, and for some time after the publication of Lyell's book, criticism continued to be made. But the concept of uniformitarianism looked so reasonable that gradually the opposition disappeared, and it became almost universally accepted.

The recognition of the principle of stratigraphic geol-

The Rise of Modern Geology

ogy gave an immediate stimulus to geologic investigation. Previous to the time of Smith, interest in the rocks of Britain had centered around the Coal Measures, or Carboniferous rocks, inasmuch as they were of the most practical importance. Above them lay a series known as the New Red Sandstone—interstratified layers of sand, clay, slate, conglomerate, and lava. Below the Carboniferous lay another series, known as the Old Red Sandstone. Below them was a mass of unclassified rocks resting on schists and other crystalline rocks, none of which was fossiliferous.

In 1822 Adam Sedgwick, professor of geology at Cambridge University, began studies on these lower rocks, and eventually extended his studies to all of western England as well as of Wales. The same time Sir Roderick Murchison, of the Geological Survey, carried on extensive work on the same rocks. Sedgwick and Murchison differed in their interpretations, one insisting that most of the group be called Cambrian and the other that it be called Silurian. Eventually a compromise was reached by setting up a third division between the two and calling it Ordovician. Thus the three lowest groups of the Paleozoic, "old-life" rocks, were established as Cambrian, Ordovician, and Silurian. All the names are from ancient Welsh tribes.

Since the Old Red Sandstone was so abundant in Devonshire, it was renamed Devonian; and inasmuch as the New Red Sandstone was common in the province of Perm, Russia, it was renamed Permian. Rocks lying above the Permian were worked out in detail by the same methods, using fossil contents as the criteria for subdivision. The lowest, Triassic, was so called because it had three divisions on the continent; the Jurassic was named from the Jura Mountains, where it was prominent; the Cretaceous took its name from the Latin *creta* for chalk, because the chalk cliffs

Fossils, Flood, and Fire

of England were of this group. These three became the Mesozoic, or "middle-life" rocks. Above them lay another series of sediments, called the Tertiary, which were worked out as part of the Cenozoic, or "later-life" rocks. Above the Tertiary came the Pleistocene, or glacial deposits, and finally the Recent. Thus the geological column was established.

Soon after the British Paleozoic system was established, studies were carried out on a group of similar rocks in New York. The New York State Natural History Surveys ran for five years beginning in 1836, under the direction of James Hall. The New York System thus established corresponds closely to the British system. It includes a similar series running in order from the underlying crystalline rocks up to the Carboniferous, which in America was soon divided into two divisions, the Mississippian and Pennsylvanian.

As studies progressed in America, many of the formamations were found to extend for great distances to the south and west. Other formations disappeared after several miles, and were replaced by new ones in similar positions, but with sufficiently different fauna and flora to make it necessary to give them different names. Eventually it became clear that there was a definite sequence of strata over the whole continent. When any formation can be traced continuously, it retains its name; however, if it becomes impossible to follow it far, because of its being hidden beneath the surface, folded, faulted, or having its continuity broken by areas through which it is impossible to trace, then a formation appearing in the same relative position is given a new name, but is considered as equivalent in stratigraphical position.

This principle of correlation is used extensively by mining and oil geologists, and is of practical import-

The Rise of Modern Geology

ance. The degree of accuracy to which scientific oil drilling has attained is almost unbelievable to one who has not had contact with the profession.

The data which have just been presented might make it appear that there is a universal sequence of fossils which would make it possible to correlate the rocks of any part of the earth with those of England and America. But a retrospective view of the data that have been assembled from the various continents makes it seem that not all parts of the earth are alike. Great regions appear to have distinctive features that set them apart. We shall not attempt to go into detail regarding this point, except to say that the same names are generally used for the major groups, or systems, even though they do not show the same types of fossils. This is merely a matter of convenience, and is based, of course, on the theory that there has been a succession of life throughout all past time, even though living types might have differed in various parts of the world.

It is quite obvious that *if evolution were true,* and there had been long ages of deposition, the current methods of stratigraphy would make it possible to arrange the rocks in order according to the sequence of life throughout the ages. But, *supposing evolution were not true,* and there had been no long ages of time, would it then be possible to arrange the rocks in order? This question we shall consider in another chapter.

CHAPTER THREE

The Delusion of Uniformity

If it were possible to demonstrate geological processes that have been active in the past, we might check our hypotheses as to their validity. Since this cannot be done, we must remain content with a certain degree of uncertainty. However, a fair amount of knowledge can be accepted as beyond any reasonable doubt. It is upon these conclusions that any science can be built. We must, nevertheless, recognize that in such cases there must be a chance for differences of interpretation, with possible errors, especially if conclusions are based on incomplete data, or on processes which we can only extrapolate from the present, with no way of proving that such extrapolation is correct in every detail.

The great problem in geology is whether geological phenomena shall be interpreted in terms of catastrophism or uniformitarianism. Diluvialism (or diluvianism, if you prefer), which is belief in a universal Flood, is based on the catastrophic hypothesis, whereas the opposite interpretation, that of long ages of time, is based on the uniformitarian hypothesis. It is the purpose of this chapter to review some of the data from geology and to ascertain, if possible, whether the

Fossils, Flood, and Fire

uniformitarian hypothesis, which today is generally accepted, is actually valid. If our investigations should leave any doubt as to its validity, then we would be obliged to grant to the diluvialists the right to their interpretation as an alternative to the popular uniformitarian geology.

When the modern theory of geology was developing during the late 18th and early 19th centuries, it was taken for granted that the ocean bottom would, if studied, show deposits of rocks similar to those on the continents. Not until the latter part of the 19th century was this notion dispelled, by the Challenger expedition sent out by the British government from 1872 to 1876. Covering much of the Atlantic, Pacific, and Antarctic oceans, the investigators discovered that the floor of the oceans is not covered by sedimentary rocks like those of the lands. Instead, microscopic shells of plants and animals, continually falling through the water, accumulate in the form of a dense, fine deposit known as *ooze*. Sometimes this is chalky, sometimes siliceous, depending on the nature of the shells. Near the shores fine sediments from the rivers form deposits of red mud and blue mud. But nowhere is anything being formed like the thousands of feet of sedimentary sandstone, shale, and limestone so well known to anyone who has studied the rocks.

Were the present sea bottoms to accumulate sediments for great lengths of time, they would probably produce thousands of feet of material, but it would consist largely of chert, a flint rock formed from siliceous remains of microscopic sea creatures. The contrast between this and ordinary sedimentary rock, which shows evidence of having been washed into place by currents of considerable volume, is very striking. And so we must conclude that the uniformitarian hypothesis falls down on this point. Wherever we look, the

Cross-bedded sandstone is evidence of rapidly changing currents. Upper: Cross-bedded sandstone in Monument Valley, Utah. Lower: Cross-bedded sandstone at Zion National Park, Utah. Photos by Ernest S. Booth.

Fossils, Flood, and Fire

picture is the same—the sedimentary rocks show evidence of extensive and vigorous water action. A few examples will illustrate this.

If we travel over the Colorado Plateau, which covers much of Utah and bordering states, we would find extensive cross-bedded shales and sandstones, with lenticular beds, laid down on the truncated surfaces below. Some of these beds are comparatively thin, sometimes from ten to a hundred feet thick, but spread out quite regularly over a vast area of 100,000 square miles or more. How such deposits could have taken place without violent and widespread waves of water is impossible to imagine. Similar conditions are found throughout India, where rapid alternation of coarse and fine sandstone indicate fluctuations in current. In the beds are thousands of trees in a state of disorder, as if dropped by rivers during a great flood. In South Africa are to be seen great masses of schist and granites which are described as having been spread out and sorted by immense currents of water. They have been likened to giant placers which worked over vast quantities of material. Certainly there is nothing of a normal, or uniformitarian nature in this.

One can hardly read any geological literature without being impressed with the many references to supposed submergences and emergences of the land surface. This comes from the great abundance of alternation of land and marine sediments. In the Paris Basin, for instance, are six cycles, the marine being from the north and the terrestrial from the south. Very likely it was this cyclic sedimentation that gave Cuvier his idea of successive catastrophies.

Another example of cyclic sedimentation is seen in the Tertiary of Burma. The beds vary in thickness from 1,000 to 12,000 feet, but the peculiar fact is that they appear to have been laid down in shallow water.

The Delusion of Uniformity

The advance and retreat of the sea can be clearly seen. But to explain this phenomenon by the theory of continuous oscillation of land levels for thousands of years is almost too fantastic to believe.

A most notable example of widespread sedimentation under violent water action is found on the Gulf coast of Texas. At least nine maximum transgressions of the sea are followed by maximum regressions. Streams coming from the interior were laden with sediments, with which they tried to build a deltaic plain. They were met by great waves that opposed them and tried to throw them back. Trunks of trees were carried downward by the currents and buried in the mud along with branches and logs. Shifting currents undercut banks of newly deposited sand and mud, and there is every indication of constant change in the direction of the currents. All these evidences contradict the basic assumption of uniformity and give support to the catastrophic hypothesis.

Many more illustrations might be given had we room to include them.

Mention might be made of the coal beds, which are generally supposed to have been produced in great bogs over a period of 50,000,000 years or more. But the bog theory is inadequate to explain conditions prevailing in the coal. The vegetable matter from which the coal was formed appears to have been washed into place by strong currents, as indicated by the nature of the roof shales and sandstones interbedded with the coal layers. Upright trees are sometimes found in coal beds, in some cases extending through several layers of coal and intervening sandstones. Sometimes trees are standing head downward. All these facts show that the material was not accumulated in a normal manner.

The exactness with which the alternating beds of coal-bearing rocks occur has caused comments by many

Coal beds alternating with sandstone, South Joggins, Nova Scotia.

observers. Successive alternations, over and over again, are characteristic of the Pennsylvanian deposits. At South Joggins, Nova Scotia, are seventy-six seams of coal; in the British coal fields as many as eighty or more beds are known, and in Germany over a hundred.

The Delusion of Uniformity

Moore, in his treatise on coal, speaks of the multiplicity of variations, accompanied by many thin sedimentary units in a cyclic rhythm of deposition, in some cases only a few feet or even only inches thick, that can be traced for hundreds of miles. Such regularity of deposit over large areas, accompanied by alternating masses of vegetation, is something unknown today. It is unanswerable evidence of conditions totally unlike anything we now have.

Evidence for violent burial of both plants and animals is the common phenomenon we find as we examine fossil-bearing rocks. A few examples will be given.

In the lower rocks, especially the Cambrian, the trilobites (see page 78) were the dominant species. These creatures, somewhat resembling sowbugs, were of many species differing from small ones an inch long to large ones over two feet in length. But of all of this vast array, not a single group has remained to this day. While so very abundant in the lower rocks, they suddenly cease to occur in rocks higher up, or fade away rapidly.

Many other animal groups show the same phenomenon, such as the echinoderms, the mollusks, and many kinds of fishes quite different from any we know today. Geologists speak of their extinction as a puzzle, a mystery, or a curious and inexplicable circumstance. This is especially true of the giant reptiles, such as the dinosaurs. This group had almost as many forms as all other land animals together. One writer calls their disappearance the most dramatic and puzzling event in the history of the earth. We have no conception, authorities say, of the cause of the extinction of this mighty race. No good reason can be found in uniformitarian geology why they should have disappeared so suddenly from the face of the earth.

What has been said of the reptiles can be said with

Fossils, Flood, and Fire

equally forceful emphasis for the mammals. Literally hundreds of extinct types are to be found in the rocks, but like the extermination of the reptiles, that of the mammals is a profound mystery to the geologists. It is, they say, an unexplained mystery. Nobody can find good reasons why so many of them should have disappeared so suddenly from the earth.

Anyone who looks toward the catastrophic interpretation, and who regards the geologic column as a series of deposits representing the burial of the life zones of the ancient world by an overwhelming cataclysm, finds in these sudden exterminations exactly what he might expect. The puzzle of the geologists becomes clear when viewed in the light of the Scripture declaration:

I will destroy them with the earth.—Gen. 6:13.

And now that we have traced the history of uniformitarian geology, and have noted a few, but only a few, of the evidences against uniformitarianism, let us turn to the alternative interpretation, that of the Genesis Flood. Let us see how it has attracted the attention of sober thinking scientists, both in the past and at the present, and let us see how it answers some of the great geological problems in a manner that uniformitarian geology cannot. In spite of the fact that this interpretation has become the butt of ridicule for the scientific world, it may possibly turn out to be a more satisfactory explanation for geological phenomena than is generally recognized. At least, it should be given a fair chance to demonstrate its possibilities.

CHAPTER FOUR

The Flood Theory of Geology

Fossils have been observed since ancient times; the Greeks and Romans regarded them as due to natural causes. The early Christians interpreted them as due to the Flood, but in the Middle Ages they were considered to be freaks of nature. Some thought they were formed in the rocks from seeds blown in from the sea. Some said they were placed in the rocks by the Devil to deceive men; others said that God placed them there to test men's faith.

Probably the first man in modern times to recognize fossils as having originated from natural causes was Leonardo da Vinci, who lived around the beginning of the 16th century. But it was John Woodward, professor of medicine at Cambridge, who convinced the world that fossils were actually of organic origin. In his *Essay Toward a Natural Theory of the Earth,* in 1695, he attempted to explain how the Flood had dissolved all the mineral constituents of the earth as well as mixing them with remains of plants and animals, and had precipitated them together, thus forming the sedimentary rocks. Though his knowledge of the rocks was meager, and his explanations full of absurdities,

37

nevertheless he followed the Bible story closely, postulating a reign of sin following the creation of man, then a period of mild climate in which man did wickedly, and then the Flood.

During the 18th century the Flood theory was universally accepted among Christians, although some scientists were skeptical and tried to bring in other explanations. John Hutchinson, in 1749, accounted for the alternation of strata by means of tidal waves flowing in different directions. Alexander Catcott, lecturer in St. John's College, Cambridge, published *A Treatise on the Deluge* in 1761, in which he pointed out that the rocks of most mountain areas must have been continuous when first laid down, and that erosion had cut down the valleys and left the mountain peaks which we now see.

He attributed most of this erosion to the period immediately following the Flood, when the strata were soft and the volume of water tremendous.

In 1799 John Williams, a mining engineer, discussed the formation of the coal beds by the Flood. He held that most of the ancient world was covered with luxuriant forests, which were swept away by gigantic waves of water and buried layer after layer with intervening layers of material that was brought from the oceans by the Flood waters.

One of the most complete discussions of the Flood theory in recent times is *The Deluge Story in Stone*, by Byron C. Nelson, published by the Augsburg Publishing House in 1931. From this notable study we quote:

> Although the Flood theory of geology was not without its assailants in the eighteenth century, and there were not wanting men to advance some other theory, the Deluge easily maintained its place as the most common explanation of the earth's geological state. The works of such men as Buffon (1749) and Hutton (1788) in behalf of the doctrine of uniformity, which modern

The Flood Theory of Geology

geology has adopted, met with little favor. The reason for this was that the leading educators of the day were largely men of great religious faith, men who believed strongly in the Bible and did not hesitate to teach others to do likewise. As the century progressed, the control of education in Europe and America passed gradually out of the hands of such men into the hands of men secretly or even openly hostile to the Bible. The result was that the theory of uniformity, which had come down from the ancients and had been advocated to some extent in the sixteenth and seventeenth centuries, gained the upper hand, and the Flood theory fell into the background.— p. 83.

George McCready Price, in his book, *The Modern Flood Theory of Geology* (1935), made the following commentary on the transition to uniformitarianism:

> The Flood theory was betrayed in the house of its friends. . . . Baron Cuvier, the French scientist and the most eminent scientific investigator and writer of his day, . . . taught the world to believe in a long series of successive catastrophes; and Charles Lyell . . . induced the world to discard an absurd and overworked Catastrophism, and to adopt a system of trying to interpret all the past changes of the globe in terms of the present, ordinary operations of nature in the world of today.— p. 33.

We have already shown how the theory of uniformity came to be quite generally accepted during the early years of the 19th century. Thus the Flood theory went into eclipse that lasted for approximately half a century. But it did not die without some men making heroic efforts to save it. As many as eight to ten books appeared during this period of decline, in which the authors tried to apply the rapidly developing knowledge of geology to the catastrophic interpretation.

The demise of the Flood theory was abetted by a theological movement that had a profound influence on church doctrines in general. French rationalism

Fossils, Flood, and Fire

was working beneath the surface, and undermining the faith of the people in the literal interpretation of the Bible. Then in 1835 German Biblical criticism was "exploded" upon America, and skepticism rapidly invaded the Protestant churches. Thus they were prepared for the acceptance of uniformitarian geology, and when Darwinism was introduced in 1859, many churchmen were ready to adopt the revolutionary views.

Although many church leaders resisted Darwin, yet they had so fully accepted uniformitarian views that when biological evolution was introduced, they were at a loss to know how to meet it, and eventually succumbed to its insiduous propaganda. They had adopted virtually all the elements of uniformitarian philosophy into their theology, and since they were not following any longer the dogmatic method of settling disputed points by reference to the Bible, they found themselves unable to halt the surge of evolutionary geology that was sweeping the country. And so, by the end of the 19th century, the triumph of liberal science and theology was almost complete.

But the victory of uniformitarian geology was not to go long unchallenged. In 1902 George McCready Price published his *Outlines of Modern Science and Modern Christianity,* in which he urged a return to primitive Christianity. Then in 1906 came his *Illogical Geology.* This created quite a disturbance in both scientific and religious circles. In this small, unpretentious pamphlet, published at his own expense, he laid down the major tenets of the Flood theory of geology, which was to be kept before the world by his writings for another half century. Two main points were made:—(1) That there is no proof of a succession of life through long ages of time, and (2) that the uniformitarian hypothesis is "unproved and unprovable." With a considerable array of technical arguments he proceeded to show the inconsistency of geology as com-

The Flood Theory of Geology

monly taught, and challenged the scientific world to consider the possibility of the Deluge story of Genesis as an alternative, in fact, the only possible, explanation for the rocks of the earth.

In 1916 came his *Back to the Bible,* a powerful arraignment of the growing apostasy into which the Christian church was fast drifting in its acceptance of evolutionary geology. Then his masterpiece, *The New Geology,* in 1923, attempted to organize all geological knowledge into a textbook to be used in colleges. So successful was it that for many years it was used in a number of Christian colleges, and an unusually large number of copies were sold, many to people who were interested to learn what catastrophic geology had to offer.

A number of other writers followed Price's lead, and for many years books and magazines echoed his ideas on geology. During the years 1920 to 1945 "Flood geology" played a large part in the thinking of theologians as well as scientists, even though the latter generally ridiculed it bitterly.

In 1920 Professor Price taught a course in geology at Pacific Union College, Angwin, California — his first adventure in college classwork. He continued to teach courses in geology and creationism in various colleges for many years. It was my privilege to be a student in his class at Pacific Union College, and for twenty-five years thereafter I conducted the course in connection with my work in the biology department. As was inevitable, the new venture into interpretation of a well-established science met with some problems, and during the years a few revisions were found necessary. When I published my *New Diluvialism* in 1946, I incorporated some of them into that work. The *New Geology* had given little attention to glaciation, and I expanded on that line, and tried to show how the data regarding glacial action could be fitted into

Fossils, Flood, and Fire

the Flood theory. Then, too, I found that there was much more regularity to the stratified rocks than Price had recognized, and this, too, was developed by explaining this order and system as due to the burial of the ancient life zones rather than to a succession of life during long geological ages. (Further details of this viewpoint will be discussed in a later chapter.) And finally, I found that there seemed to be clear evidence for extensive lateral movements, known as *overthrusts*—a point which had hitherto not been recognized by diluvialists.

One point should be emphasized emphatically, with respect to these new interpretations, and that is, that none of them in any way changed the basic premise of Price's "Flood geology," that the Flood was a universal catastrophe occurring, according to Biblical chronology, not many hundred years before the beginning of written history, and that the major features of geology must be attributed to this Deluge. Price and I have both held rigidly to the thesis that there was no life of any kind anywhere on the earth before the third day of creation week, which according to Genesis chronology, occurred about 6000 years ago. Any other interpretation, we have maintained, is the result of thinking in terms of uniformitarian geology, which we have held to be false.

In 1960 a symposium on geology was held in Loma Linda, California, under the auspices of the quadrennial conference of Seventh-day Adventist science teachers, and at this time suggestions were made that part, at least, of the Tertiary rocks may have been produced after the Flood. This possibility has been studied carefully, and seems to have some merit. It will be discussed later. Since then studies on Lower Paleozoic rocks seem to indicate that some reefs may well have had their origin in the deep seas before the Flood, and this seems quite acceptable also. The problem of the

The Flood Theory of Geology

reefs—to what degree they were changed by the Flood, to what degree they may have been buried *in situ,* or have been transported—such points as these are subjects that will demand much study before definite conclusions can be reached as to how much geological action may have occurred before the Flood.

In 1961 a new book on the subject of "Flood geology" appeared, *The Genesis Flood,* by Whitcomb and Morris (Presbyterian and Reformed Publishing Company, Nutley, N. J.) John C. Whitcomb, Professor of Old Testament, Grace Theological Seminary, Winona, Indiana, wrote the first portion, dealing with the nonscientific aspects of the Flood theory. Henry M. Morris, Professor and Head of the Department of Civil Engineering, Virginia Polytechnic Institute, Blacksburg, Virginia, wrote the geological part. Morris is a scientist of high repute, and his ideas cannot be lightly regarded. He presents many points against the commonly accepted uniformitarian geology. Even though we might not agree with him on every detail, nevertheless his ideas are stimulating, and are worthy of careful consideration. They constitute a powerful argument for the Flood theory of geology.

In 1963 Morris published another volume, *The Twilight of Evolution,* in which he contended that evolution is declining in status with men whose ideas are centered in the Bible record of creation. Among those who have not been sold on the evolutionary theory are scientists and geologists of stature. He states that the entire concept of evolution is not only non-Biblical, but is also non-scientific. He devotes one whole chapter to the history of geology. Here is a notable statement:

> The fossils *must* have been laid down under sudden and probably catastrophic conditions, or else they would not have been preserved as fossils at all. Even such a consistent evolutionary uniformitarian geologist as Dunbar

Fossils, Flood, and Fire

recognized that practically all fossils must have been formed by flood or other catastrophes.—p. 62.

We might, if space allowed, mention other influences that are being felt in opposition to the popular views of geology. Only a few can be mentioned. The GeoScience Research Institute set up in affiliation with Andrews University at Berrien Springs, Michigan, is doing extensive studies on the problems of geology and the Genesis record. At present writing, no publications of note have appeared. Recently a non-denominational organization, The Creation Research Society, which has a membership of around 500, all of them scientists of good standing, has been organized. One of their main articles is that to become a member one must accept literally the Genesis record of the Flood. Another organization, the Bible-Science Association, holds to the same basic premise. Other organizations might be mentioned, but suffice it to say that "Flood geology" is undergoing a resurgence, and the fundamental assumptions of uniformitarian geology are being challenged more than they have been for a century. The situation is fluid, and will bear watching; no one can tell what the future will bring forth.

CHAPTER FIVE

Flood Legends

Preserved in the myths and legends of almost every people on the face of the globe is the memory of the great catastrophe. While myths may not have any scientific value, yet they are significant in indicating the fact that an impression was left in the minds of the races of mankind that could not be erased.

> There never was a myth without a meaning; . . . there is not one of these stories, no matter how silly or absurd, which was not founded on fact.—Bancroft, H. H. *Works*, Vol. III—*The Native Races of the Pacific Slope; Mythology.*

The Assyrians and Babylonians had several Deluge legends, of which the Epic of Gilgamesh is the closest to the Biblical account. It, as well as other legends from the Mesopotamian valley, speaks of a man commanded to build a ship and stock it with every living thing. Cattle and wild beasts were sent to the ship and the door was shut. Then the rain came, accompanied by lightning, thunder, earthquakes, torrents from subterranean channels, and darkness. Finally the storm abated, the ship grounded, and the man sent forth a dove, a swallow, and a raven. When the man left the

Fossils, Flood, and Fire

ship he offered sacrifices, and the gods spread a rainbow in token of their promise not to forget him.

It is claimed by some scholars that the Hebrew story of the Flood was derived from the Epic of Gilgamesh. However, the absurdity of such a claim is apparent when we examine the two records in detail. The Genesis story is that of universal destruction brought upon the earth because of its wickedness, and the care of Jehovah for those who trusted in Him. It is reasonable and consistent all the way through. On the other hand the Assyrian legend is so full of inconsistencies that it makes no sense to the believer in the true God. Let us notice a few of the inconsistencies.

Instead of punishment for the sins of men, as the Genesis story implies, the Deluge, according to the Assyrian myth, was the result of controversy and jealousy among the gods. How little they cared for man is indicated by the fact that they did not warn him of the coming destruction, but kept it a secret. The Babylonian account only allows six days of rain, and fails to mention other causes; this of itself is not sufficient cause for a world-wide event of the magnitude of the Flood. Then in the Babylonian account the gods quarreled over the sacrifice that the man offered them; finally one of them took the part of man and exalted him to be a god. The whole legend is shot full of false religion, and cannot be accepted as of any value except as merely another legend that shows how strongly the events of the Flood lingered in the memory of the race.

The Greeks celebrated yearly a day of mourning for the dead who were victims of the great Deluge. They had several legends which quite closely resembled the Genesis story of the Ark.

The traditions of the ancient Britons tell of a time when mankind had become so wicked that a tempest of wind, fire, and flood was sent to the earth. However,

Flood Legends

one man, distinguished for his integrity, was shut up in an inclosure with a strong door until the storm abated.

Hesiod, an early Greek poet, tells of a great catastrophe that brought thunder, lightning, and floods of water. Huge billows, he said, roared around the shores. In places the earth melted, and raging fires poured forth.

The Latin poet Ovid, in the first century B. C., wrote a description of a flood which corresponds closely with the Biblical record.

Norse legends tell of a flood, with ice and fire, brought upon the earth as the result of the evil deeds of men. The land extending westward was destroyed, and after the flood came a long and cruel winter. This continued until a layer of ice formed all the way across to Greenland.

Flood legends have been noted among many of the Asiatic tribesmen. All across Siberia are tales of a man who built himself a great ship, into which he gathered specimens of animals. The flood was so great that it destroyed all the animals outside the ship; bones of animals found in the earth were regarded as evidence of the great catastrophe. In the tale of the Sagaryes the man's name was Noj. The Altaic tale calls him Nama, and is like the Biblical story, almost identical in all details.

Many of the stories tell of fire as well as water. Evidently this is due to a remembrance of vulcanism that accompanied the Flood.

In North America the Eskimos have a Deluge story; and it is universal among the American Indians. In all their stories water and fire are mentioned, and often a period of cold comes after the flood.

In most of the legends of Latin America the early world was overwhelmed by fire and flood in successive catastrophes. The Quiches of Central America tell of

Fossils, Flood, and Fire

a race of men that had intelligence but no moral sense, so they were destroyed by fire and pitch from heaven, accompanied by terrific earthquakes. The eruptions were accompanied by great rains. Because of these conditions famines developed, as the earth was too cold and moist to produce food.

The Fijian legends tell of a rain that submerged the earth, whereupon eight persons were saved in a canoe. The Hawaiians say the wicked men were all destroyed but Nu-u. He escaped in a great canoe stocked with food, plants, and animals. The rainbow story is connected with this legend.

Going back to the Middle East again, we find the Sumerian account quite interesting. It tells of two periods, one before the great flood, one after. Before the flood ten kings ruled. However, each one is supposed to have reigned for a period of thousands of years— possibly a remnant of the Biblical record of the unusual age of the patriarchs. While the legend confines the flood to a small portion of the lower Mesopotamian valley, yet it follows the Biblical story fairly well by telling of a man who was saved in a great boat.

The Encyclopedia Britannica, while strongly modernistic and critical in its treatment of flood legends, makes the interesting comment that "it is quite insufficient to dismiss these stories as historically worthless." (Art.: Genesis)

Of all the legends, those of the Chinese are perhaps the most striking. They tell of a period of wealth and happiness, after which men departed from the ways of righteousness, and the earth was destroyed. The stars of heaven were moved out of their places, apparently by the rocking of the earth.

Byron C. Nelson, in his *Deluge Story in Stone,* reports that on a Buddhist temple in China a traveler reported having seen, in a beautiful stucco a painting where the goddess of mercy looked down from heaven

Fossils, Flood, and Fire

upon the lonely Noah in the Ark, surrounded by raging waters. Dolphins were swimming around him, and a dove, with an olive branch in her beak, was flying toward the vessel.

Only a few of the many myths and legends of the great Deluge have been cited, and these have been mentioned with great brevity. But what has been given is sufficient to show that there is a universal memory of a terrible catastrophe that is supposed to have destroyed the human race except the few who were miraculously preserved, and that their descendants have kept its memory alive ever in traditional tales. Certainly such a universal prevalence of Flood legends could not be possible unless there had been reality upon which they could have been founded.

CHAPTER SIX

Ancient Life Zones

It is of considerable interest, and possibly of no small importance to note that when Charles Lyell published his uniformitarian views in 1830, his ideas were not wholly acceptable to the geologists. In 1831, when Adam Sedgwick retired from the presidency of the Geological Society, he gave a long address reviewing the new views. In this discourse he criticized strongly the uniformitarian theory. He maintained that the same principles of animal distribution that are recognized in modern times must have applied in ancient times.

A few years later, in 1859, Herbert Spencer, although an ardent advocate of evolution, attacked the uniformitarian theory vigorously. Werner's "onion-coat theory" of the regular succession of lithological deposits was illogical, he declared; but to make matters worse, current geological theory had adopted a biological onion-coat theory that assumed that similar forms of life had lived simultaneously in different parts of the world, and that in any certain period of earth's history only certain types were in existence. The fact that the fossils changed from place to place indicated, he argued, that, as Sedgwick had previously stated, different

Showing major events and characteristic life of each system

Classification			Major Events	Characteristic Animals	Characteristic Plants
CENOZOIC	RECENT		Present conditions developed	Modern man, plants, and animals	Deciduous trees
	PLEISTOCENE		Vigorous erosion and glaciation	"Primitive" man; mastodons and mammoths	A few flowering plants
	TERTIARY	PLIOCENE	Rise of great mountain systems	Many modern plants and animals	
		MIOCENE	Intense vulcanism	Transition to modern	
		OLIGOCENE	Vigorous erosion and vulcanism		
		EOCENE	Continued mountain building		
		PALEOCENE	Survival of some forms from lower zones		
MESOZOIC	CRETACEOUS		Great submergence of newly formed continents; formation of chalk, coal, and oil.	Giant dinosaurs Marsupials and a few placental mammals	Cycads dominant; trees, ferns, conifers
	JURASSIC		Uplift of Rockies Uplift of Sierra Nevada Formation of "Red Beds"	Many reptiles; modern insects; ammonites	
	TRIASSIC		Complex tilting, folding, and erosion of Appalachians	Fresh-water fishes; small dinosaurs; reptiles; ammonites abundant	Conifers, cycads, tree ferns, scouring rushes
PALEOZOIC	PERMIAN		Widespread tectonic action	Many small reptiles	Conifers; ferns
	CARBONIFEROUS	PENNSYLVANIAN	Formation of coal beds	Many fishes, reptiles; few insects; crinoids	Scale trees; tree ferns; few seed plants
		MISSISSIPPIAN			
	DEVONIAN		Appalachian uplift	Armored fishes; a few amphibians and insects	Ferns, tree ferns, horsetails, mosses
	SILURIAN			Mollusks; corals	A few simple land plants
	ORDOVICIAN		Deposition of limestones, marbles, slates. Tectonic movements in many places	Marine invertebrates; trilobites dominant; a few fishes	None known
	CAMBRIAN		Beginning of erosion of lands	Marine invertebrates; trilobites, brachiopods	Calcareous marine algae
PRE-CAMBRIAN	ALGONKIAN		Deposits of shales and conglomerates	Worm tubes	Algae?
	ARCHAEAN		Primitive rocks, with many intrusions of lava and granite; much distortion	No plant or animal fossils known	

Ancient Life Zones

assemblages of life had existed together even as they do today.

The challenge to geological theory presented by Sedgwick and by Spencer seemed to have had little effect. There were some, however, who did not accept the popular views, but who thought that the catastrophic theory of geology could be maintained scientifically. In 1906 George McCready Price published his *Illogical Geology,* in which he expanded upon the objections Spencer had raised, and pointed out the inconsistency of the whole evolutionary theory. It is possible, he said, that

> some type of fossil might occur next to the Archaean in Wales, and another type in Scotland, while still another type altogether might be found in some other locality, and so on over the world, leading us to the very natural conclusion that in olden times there were *zoological provinces* and districts.—p. 64.

Thus, he maintained, there was no such thing as a succession of one type of life after another, but they might all have been simultaneous, and therefore the geological formations simply represent a taxonomic or classification series of the ancient world.

As Price developed his views on catastrophic geology during the early years of the twentieth century, he felt that this principle of the simultaneous existence of the various assemblages of animals and plants forbade the possibility of any sequential arrangement of the fossils. On the other hand, he ardently argued, any attempt to arrange them in order was purely arbitrary and without actual basis in fact.

However, as oil geology and mining geology extended rapidly, more and more detailed sections of the rocks were worked out by the field geologists. I remember talking with one of them and discussing this point. His answer to my inquiries was something to this effect: I don't care what theory you hold regarding

Fossils, Flood, and Fire

STRATIGRAPHY OF THE
WIND RIVER MOUNTAINS, WYOMING

Pleistocene	Glacial moraines
Tertiary	Alluvial, sand, silt, clay, and pebbles, volcanic ash; some mammal and plant fossils.
Cretaceous	
Dakota	Sandstone and shales, with fish scales, oysters, gastropods; some coal.
Jurassic	
Morrison	Clay, silt, sandstone.
Sundance	Limestone, sandstone, shale.
Triassic	
Chugwater	Siltstone, claystone, conglomerate.
Permian	
Phosphoria	Detrital beds, chert, sandstone, rock phosphate.
Pennsylvanian	
Tensleep	Sandstone, limestone; typical brachiopods, gastropods, foraminifera, conodonts, and other fossils.
Mississippian	
Sacajawea	Red shales and cherty limestones.
Madison	Limestone and dolomite; abundant brachiopods, crinoids, and corals.
Devonian	
Darby	Very little dolomite.
Silurian	None present.
Ordovician	
Bighorn	Dolomite, with few fossils.
Lander	Sandstone; extensive fauna of Orthoceros, typical Ordovician fauna.
Cambrian	
Gallatin	Limestone and conglomerate; brachiopods, crustaceans, trilobites, cystoids abundant.
Gros Ventre	Micaceous shale and sandstone; fossils rare, but those present are typical Middle Cambrian.
Flathead	Sandstone; fossils rare.
Pre-Cambrian	Graniodiorite, schist, and quartzite.

Ancient Life Zones

the origin of the earth. I find certain fossils in certain order, and that is all I know or care.

The result of these extended studies was that by the middle of the century the validity of the sequence of formations has become established beyond any serious question. Whereas it had been supposed that nowhere could more than two or three "periods" be found in order, it was now seen that in some places practically the whole geological scale could be seen at once. Such an example may be found in the canyons of the Wind River mountains in Wyoming. On the Colorado Plateau, while not all the systems can be seen in any one locality, they can be traced as they overlap from the Uinta mountains to the Grand Canyon, in correct sequence. In many places such sequences can be followed.

With this new aspect of the problem, diluvialists have been led to examine more critically their concept of the zoological provinces, habitats, or life zones —call them what you will—of the ancient world.

Let us take an overall survey of the geological series, and see what picture they present. The lower systems, from Cambrian to Silurian, are practically, if not entirely, marine. It is evident that they represent the life of the sea bottoms. The Devonian has an abundance of fishes, and is practically all marine, although considerable plant life, such as might have lived along the shores, shows up in this system.

The Mississippian, while largely marine, has some plant and animal life of the lands. Then the Pennsylvanian is a system of immense forests of trees and other vegetation types, with amphibians prominent. The Permian is similar to the Pennsylvanian.

The Mesozoic rocks contain an abundance of plants, but they are largely of a different nature than those of the Pennsylvanian. Reptiles are among the dominant animal types. Many of the sea creatures are differ-

Paleozoic strata of the Grand Canyon. Upper: view from the North Rim; lower: view from the South Rim. Photos by Ernest S. Booth.

Ancient Life Zones

ent, also. In the Tertiary the mammals are prominent, as are plants that are much like modern types. Now if we examine one of the elaborate tables of correlation that are worked out for geological formations, attempting to place them in their relative chronological position, we can see a close resemblance to what we might work out for living types in an ecological sequence. As an illustration, let us take the forest zone of the Sierra Nevada mountains in California. In the lower belt are found the digger pine and blue oak, above that the yellow pine and black oak, then the white fir and incense cedar. If we go to the Rockies we find that the middle zone, with climate similar to the same zone in the Sierra Nevada, has the same typical tree, the yellow pine. The woodland belt below it is characterized by the scrub pine, which grows in conditions like those preferred by the digger pine of the Sierra Nevadas. Above the yellow pine is the spruce, which corresponds to the white fir of the Sierras. Thus we can correlate the forest belts, using the yellow pine that is common to both, and finding equivalent climatic, or ecological types in both situations. This same principle can be applied in many other localities in the world, and is true in the sea as well as on the land.

Now this is exactly the same situation as is found in the rocks. Why, then, might we not believe that these zonal conditions prevailing in the rocks were due to environmental factors rather than to a succession in time?

Let us imagine a flood of waters rising higher and higher on the slopes of the Sierra Nevadas, tearing into the soil and washing away the vegetation and spreading it across the valley below. As the waves rose higher and higher, they would deposit layer after layer in the valley, in the same order that the plants grew in the mountains. Thus, if in after years these

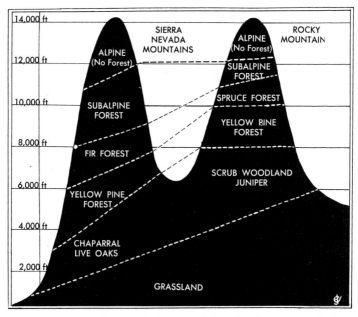

Correlation of Life Zones in the Sierras and Rockies. Note that the Yellow Pine zone is common to both. We assume that the Fir Forest of the Sierras correlates with the Spruce Forest of the Rockies, and the Blue Oak-Chaparral of the Sierras correlates with the Scrub zone in the Rockies. This indicates how geological correlation could be interpreted to indicate original zonation.

deposits were examined, they would show the succession of life zones existing in the mountains. This order of succession we might designate as *ecological zonation*, meaning a series of zones indicating the original ecological, or habitat relationships.

This interpretation, then, is suggested as a substitute for the commonly accepted theory of geological ages. In other words, an "age" of time would be replaced by a "stage" of Flood action.

Some cautions must be observed, however, and we shall point out a few.

Ancient Life Zones

Obviously in a series of deposits produced by cataclysmic action, we could not expect to have the correlation between the sequence in the rocks and the original positions of the living types absolutely exact. The violent wave action and the action of strong currents would, naturally, produce some degree of mixing of types.

Some types that might have lived in certain habitats may have been able, because of their mobility, to escape to higher levels before they were overwhelmed. Thus their relative position might not be correctly represented. Such would probably be the case with mammals as contrasted with amphibians or reptiles, which would not be able to escape so readily.

We must not suppose that all types of life would be buried *in situ,* that is, right where they lived. The violent currents that have apparently been active in producing the deposits would carry some forms for a long distance before dropping them.

And we must not suppose that all the water habitats of the ancient world were at what we might choose to call "sea level." It is perfectly reasonable to assume that there might have been immense bodies of water at different elevations. Perhaps this would account for the changes in certain water dwellers, such as mollusks, that are found in different "ages," that is, in different systems.

For these reasons, and possibly for other problems that might be suggested, the ecological zonation theory does not pretend to give an absolutely true picture of the relative positions of all life forms in their original places. But it does propose that what sequence there is, is due to the burial of ancient life zones or habitats that lived contemporaneously, and not to the succession of life throughout long ages of time.

Fossils, Flood, and Fire

The concept is simple, in fact so simple in its primary aspects that some may find it difficult to grasp. But its very simplicity makes it all the more reasonable. It is merely a question of *area* rather than *time*.

PART TWO

The Rocks and the Flood

We have now traced the history of the two conflicting philosophies by which the rocks have been interpreted: (1) the uniformitarian hypothesis, which is generally accepted today, and (2) diluvialism, or diluvianism if you prefer, or as sometimes designated, "Flood Geology." It has been pointed out that although the latter view was driven off the stage by the rising tide of uniformitarianism, it is now resurgent and militant, even though the number of its supporters may be comparatively small.

Now we intend to analyze geological phenomena from the viewpoint of diluvialism, taking a positive rather than a negative approach. The reader should be reminded that all accounts of happenings in the prehistoric past are more or less speculative, usually being of the nature of deductions from observable data. And while the popular geological discussion of past ages is generally looked upon as scientific, we must remember that it is only an interpretation and

Fossil tree trunk in upright position, South Joggins, Nova Scotia.

Part Two—The Rocks and the Flood

not absolute fact. Therefore we have as good a right as anyone to suggest an interpretation.

In attempting an interpretation of geological data one must face certain problems, and one of the greatest is that for around 150 years the uniformitarian hypothesis has been used to interpret all geological data. Thus, in reading geological literature we find it extremely difficult to separate the facts from the theories. As an illustration, we might cite the petrified trees in the Joggins beds in Nova Scotia. When Lyell made his report on these beds about 125 years ago, he viewed them from the angle of slow, uniform deposition, and made all his interpretations accordingly. Since then practically everyone who has examined them has followed his lead and has regarded them in the same way. It takes a good deal of courage and independence of thought to break away from a popular interpretation and make one that differs from it. Most investigators are afraid of being regarded as extremists or incompetent if they think for themselves. And so the long-established lines of interpretation have been accepted, generally without their validity being challenged. It is this way with many others areas—there is too much tendency to follow "authority" instead of standing on our own feet and drawing our own conclusions.

Incidentally, we may note that recent studies regarding these beds indicate that burial by violent water action is far more plausible than the commonly accepted notion of *in situ* growth.

But, someone may say, how can we be sure we are right? Perhaps we may make mistakes, and then look at how the scientific world will laugh at us. Suppose they do, it will not be the first time an interpretation has been ridiculed, even among the so-called experts themselves. All science makes progress by setting up hypotheses and then testing them. It is the right of

Fossils, Flood, and Fire

the diluvialist to set up the Flood hypothesis and then describe geological phenomena in harmony with it. This does not mean that all interpretations will be correct; nevertheless, it is the only way a new idea can make its way. And so we need not be disturbed if in the attempt to orient geology with the Flood record of Genesis we may run into some problems.

One principle should ever be kept in the mind, and that is that when a problem is met, we must not abandon previously established positions simply because we find something we are unable to explain. In such a case it is wise to lay the problem aside temporarily, until more light on it is available. In time many of the perplexing questions will become clear, if we can gather sufficient data regarding them.

And so, with these points in mind, let us enter upon a discussion of what we hope may eventually become a valid science, the science of *Flood Geology,* and not merely a hypothesis, as some now regard it.

CHAPTER SEVEN

The Basement Complex

In tracing the record of the Flood in the rocks, we must begin at the bottom. But where is the bottom? What rocks could be considered as the original rocks upon which the Flood built up a progressive series of deposits?

The earth consists of an inner core of very dense material, probably iron and nickel, then an outer mantle, and finally a thin crust on the surface. The term *crust* does not imply a solid shell over a molten interior, but is simply a holdover from the days when the interior of the earth was believed to be molten.

Science has absolutely no information as to how this great core of iron and nickel, with its covering of rock, could have come into existence. Various hypotheses have been propounded, but none of them has any scientific validity; they are pure speculation. The simple Biblical declaration that "He spake and it was; he commanded and it stood fast," is the only authoritative statement we have regarding the origin of the earth.

The crust varies in thickness. Beneath the oceans it is believed to consist of about three miles of basaltic rocks. These are heavy; in fact, the geologists believe

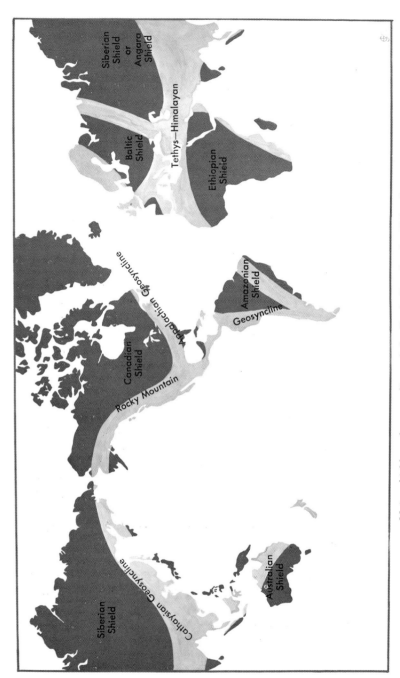

Major shields and geosynclines of the Earth. Drawing by Elden James.

The Basement Complex

that the density of the rocks has been an important factor in causing the ocean bottoms to sink down. On the continents the lighter granitic rocks vary in thickness, but may be as much as twenty miles thick in places. The mantle below the crust is supposed to be made of vitreous (glassy) rocks of varying composition.

The sedimentary rocks, consisting largely of limestones, sandstones, and shales, form a thin layer laid down by water on the crystalline rocks of the crust. These vary in thickness, sometimes reaching as much as 50,000 feet, but usually being less than 20,000 feet.

The crystalline rocks lying below the sediments have been called the *basement complex*. Two divisions have been made, the Algonkian above and the Archaean below. When we examine them we find many problems, for they have been so terrifically metamorphosed that it is difficult to decide just what might have been the original rocks before the Flood.

Pre-Cambrian rocks are located in what are generally spoken of as *shields*. The main ones are: Canadian Shield, Amazonian Shield, Australian Shield, Baltic Shield, Ethiopian Shield, and the Angara Shield—the last one in northeastern Asia.

The rocks of the Canadian Shield, comprising most of northeastern Canada, furnish the best example of pre-Cambrian rocks in America. These rocks lack fossils, and consist of fine crystalline material, which is very hard. Schists are common, and are made of finely laminated crystals. A schist is supposed to have been derived from shales by action of heat and pressure. If a shale is low in potassium, the rock will be dark, with a spotted appearance. If, on the other hand, there is an abundance of potassium, crystals of mica will be formed, and the rock will be a *mica schist*. The mica crystals are arranged in thin layers, giving the rock a banded appearance, which is sometimes beautiful, especially if the dark layers are intermingled with layers

Fossils, Flood, and Fire

of light-colored material, as may be seen in the rocks along the shores of Lake Superior. Here the pink and gray often make striking patterns.

Another banded rock is gneiss (pronounced: *nice*). It has the same three components as granite—quartz, feldspar, and mica or hornblende—but with the crystals distorted or drawn out into thin bands. It is generally supposed to have been produced by the metamorphosis of granite.

The *Vishnu schists* of the Grand Canyon are similar to gneiss in structure, but are much finer grained.

Both in the Canadian Shield and in the Grand Canyon, in fact, in basement complex rocks everywhere—in Scandinavia, Africa, Austrialia—the gneisses and schists have been intruded by immense masses of granite and similar rocks, coming up, of course, in molten form, and hardening. In the Lake Superior region alone it is estimated that over 20,000 cubic miles of granitic intrusions have taken place, and the same situation prevails in the shields everywhere.

Geologists, assuming that millions of years were involved in the formation of the basement complex rocks, postulate the wearing down of vast mountain systems to form what are called *peneplains* (*pene*: almost), regions that have been eroded almost to plains. This idea is based on the fact that although the pre-Cambrian rocks have been intensely folded, yet the surface seems to have been leveled off, rather than having been eroded into canyons such as we see today in areas of intense erosion. Diluvialists have suggested that this may be an evidence of the great forces at work on the first two or three days of creation, as the mass of the earth's material was being molded into shape.

Localities where pre-Cambrian rocks are exposed, besides those of the shields, are so rare that we are not warranted in drawing any positive conclusions as to

The Basement Complex

the manner of their formation. In the Grand Canyon the Archaean rocks do show a moderately flat surface upon which the Algonkian deposits have been laid down. But in the Canadian Shield there are so few deep cuts that it is hard to study the contact between the two. We cannot judge the nature of the pre-Cambrian by examination of the crystalline cores of mountain regions, because most of the great mountain cores, as in the Rockies, have been pushed up after the sediments were laid down. These granitic cores date from much later, relatively, geological sequence than do the granites of the shields, whether we think in terms of millions of years of geological time or the short period of the Flood. In other words, the cores of the mountains must be kept distinct from the basement complex of the shields.

The Algonkian rocks vary considerably from the Archaean. In the region of Glacier National Park the Beltian System occurs, with limestone, sandstone, quartzite, and shale, which are largely sedimentary in origin. Lava flows are interbedded with the sediments. A few fossils interpreted as algae and worm tubes, have been found, although the nature of these is not by any means certain. The sediments are from 12,000 to 25,000 feet thick, and vary considerably in the degree of metamorphism. The distinction between the Belt series and the overlying Cambrian is difficult to maintain, due to the fact that metamorphic processes produced by vulcanism have destroyed the texture of the lower rocks and tectonic forces have affected the upper and lower rocks alike. The lithological characters are similar. The distinction is generally made on whether or not the rocks contain fossils. In some areas certain rocks have been shifted back and forth from one classification to another. In some instances the Cambrian rests directly on the Archaean. This is generally true also of the Canadian Shield, and it is practically impos-

Fossils, Flood, and Fire

sible to correlate the rocks of one area with those of another.

Lying above the Archaean rocks of much of the Canadian Shield are thousands of feet of conglomerates—rocks formed by cementation together of fragments that have been broken from other rocks and rolled about in water until they have been rounded. The evidence points to violent waves that wore away the upper surface of the region and deposited these conglomerates around the southern and western margins. Masses of metamorphosed marble, slate, and quartzites may be found up to 15,000 feet in thickness. Extensive masses of rocks commonly interpreted as lavas are interbedded with them. How these lavas came about is extremely difficult to imagine.

The Blue Ridge Mountains and the Atlantic coastal plain consist of the same type of crystalline rocks as we have noted in the Canadian Shield. These are apparently remnants of a great continental mass that lay out in the region now occupied by the western Atlantic. Earth movements caused the western edge of this mass to be thrust upwards, forming the Blue Ridge, and in so doing, to exert terrific pressure on the sediments in the trough to the west, throwing them into great folds.

The question naturally arises: How can these pre-Cambrian basement complex rocks be fitted into the Flood picture? Were they formed during creation week or at the time of the Flood? These questions present many difficult problems, and at the present time our state of knowledge will not allow us to formulate a positive answer. But possibly a few suggestions can be made.

For one thing, the high degree of metamorphism, the igneous intrusions, and the confused layers of the pre-Cambrian rocks clearly indicate terrific disturbances. How much of this action we can refer to Creation Week is not easy to say. But it does seem quite reasonable that

Folded strata at Sheep Mountain, Wyoming. Photo by Ernest S. Booth.

the solid mass of dark gray rock that does not show much evidence of disturbance could be readily interpreted as the original rock of the crust of the earth.

But what about the highly distorted and metamorphosed Archaean rocks and the granitic intrusions? It would hardly seem that these effects could be thought of as being involved in the creation process, but that they would more reasonably be considered as due to the action of the Flood.

When we study the Algonkian rocks, the puzzle deepens. They are so clearly of a sedimentary nature, even though they contain no fossils, that the suggestion has been made that they must have been produced by the Flood. However, there are difficulties in this interpretation. In the first place, since they seem to underlie the fossiliferous rocks everywhere, how could we explain their being laid down there without involving any fossil-producing organism? Then, too, many of the lower fossiliferous rocks, such as Cambrian, Ordovician,

Folded, tilted and twisted strata in the Canadian Rockies, Highwood Pass area of the Kananaskis District, east of Banff National Park, Alberta. Photos by Ernest S. Booth.

Folded, tilted and twisted strata in the Canadian Rockies, Highwood Pass area of the Kananaskis District, east of Banff National Park, Alberta. Photos by Ernest S. Booth.

Fossils, Flood, and Fire

and Silurian, of the Midwest, show little evidence of violent action, but appear to have accumulated under comparatively quiet conditions, perhaps before the Flood. The nature of the Algonkian rocks and the lower Paleozoic rocks is so different that it seems difficult to explain them as due to the same process.

Might it not be possible that when the basement rocks, the Archaean, were first formed, certain movements of water over them caused their surfaces to be eroded and the materials to be swept off and spread about widely? If so, when later on, the lowest fossiliferous rocks were deposited, they would rest naturally on this primordial base. Then, when the violent actions of the Flood took place, the intense metamorphism would produce the gneisses and the schists, and bring up intrusives from deep down in the earth. The whole picture is an obscure one, and one that needs much study, but we suggest it as something worthy of attention. Diluvialists need to grapple with these problems and develop a real positive approach to the technical questions involved in them.

In spite of the uncertainties we may find in regard to these rocks, one thing should not be overlooked. They do give evidence of most astonishing degrees of metamorphism. In the Appalachian region the Paleozoic and Algonkian rocks are folded and metamorphosed until they cannot readily be distinguished in many places.

Speaking of the problem of the pre-Cambrian of the Canadian Shield, one recent report states:

> The application of rock and time terms to the pre-Cambrian strata has become much confused. . . . Standardization of nomenclature is nearly non-existent and it is practically impossible to obtain any general acceptance for the meaning or usage of particular terms.—*Geology and Economic Minerals of Canada*, p. 25. (Geol. Surv. of Canada, Econ. Geol. Ser. no. 1, 1957)

The Basement Complex

This statement shows plainly that these rocks afford a real problem for the geologists, and we need not be too concerned if we find in them some hard questions to answer. We must recognize that a great deal more must be learned about these rocks before positive conclusions can be reached.

Tilted strata of limestone at Lake Titicaca, Peru. Photo by Ernest S. Booth.

CHAPTER EIGHT

Life of the Ancient Seas

We now come to the fossiliferous rocks, and in this chapter we will deal with the Cambrian, Ordovician, Silurian, and Devonian strata, which, with the exception of the Devonian, contain little but marine life. We shall note the relation between these groups of rocks and those immediately above them, which show some types of land life. The rocks of the lower divisions are much alike, consisting largely of sandstones, shales, and limestones, and are prominently displayed in the major geosynclines* and nearby areas over the whole world.

The most striking feature of these Lower Paleozoic rocks is the complexity and abundance of the animal life in contrast to the almost complete absence of life in the Algonkian rocks immediately below them. This has led to a great deal of speculation. How could such a rich and varied assemblage of life have come into being without having left any previous record, if it

---

* *Geosyncline.* All over the globe we find evidences for the existence in the past of long, narrow, and comparatively shallow waterways. These have been filled with sediments, and afterwards have been upheaved to form the main mountain ranges of the world. Because they are world-wide in extent, they are called *geo* (earth) *synclines* (basins). Examples are the Appalachian, the Alpine, the Himalayan, and Rocky Mountain geosynclines. (See p. 66)

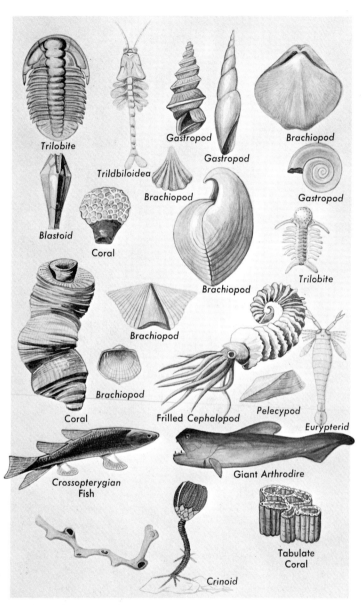

A group of typical Paleozoic fossils. Drawing by Elden James.

Life of the Ancient Seas

had come by evolution? Geologists imagine that life must have existed for at least a million years, possibly several million, in order to have become so abundant and complex, yet nowhere can any primitive fossils be found that could be accepted as precursors of the lower Cambrian life.

Many of the pre-Cambrian rocks are as little metamorphosed as the ones above them, yet they show no fossils. Therefore it cannot be said that the fossils were destroyed by metamorphism. It is apparent that they never did have any fossils; this is a great puzzle to the geologists.

As an illustration of the Cambrian fauna we might mention the *Olenellus fauna,* as it is called, after the trilobite *Olenellus.* This fauna is world-wide in extent, and is considered the lowest stratigraphically of any Paleozoic animals. It consists of sponges, jellyfishes, corals, starfishes, worms, brachiopods, bivalves, and trilobites.

Perhaps the most important Cambrian fauna ever discovered is that of a bed of black shale near Field, British Columbia. The Burgess shale fauna contains the remains of many soft-bodied animals pressed flat like the flowers in a press, and perfectly preserved. As many as 130 species have been described from a bed only a few feet thick. Most fossils consist of creatures with shells or bony skeletons, but this one is remarkable because of the preservation of so many soft creatures. The indications are that they must have been buried rather suddenly or they would have disintegrated.

In all of North America more than 1200 different kinds of animals have been found in the Cambrian strata, representing all the major phyla except the vertebrates. Trilobites make up about 60%. They were mud-grubbers, that is, they gathered their food

79

Famous Trilobite quarry on the slopes of Mount Stephen, Yoho National Park, British Columbia. Photos by Ernest S. Booth.

Life of the Ancient Seas

from the muddy sea-bottoms. Next in importance were the brachiopods. These resembled two saucers fitted together face to face, and were attached to the bottom by stalks.

When we study the Ordovician strata we find them to be much like the Cambrian, with a few types such as graptolites, corals, crinoids, bryozoa, and clams either new or greatly increased in numbers. The graptolites resemble Chinese characters in appearance; they are related to the corals, but look like a sprig of a plant with branches sticking out abruptly. Crinoids have been called "sea-urchins on a stalk" or "sea-lilies." They were so abundant that great masses of Silurian limestones have been found to consist almost entirely of their skeletons. Bryozoa—"moss-animals"—are somewhat related to the corals also.

One of the most outstanding features of these lower Paleozoic rocks is the presence of extensive reefs. For instance, in the Niagaran formation of the Silurian these reefs extend from the Arctic to southern Illinois and as far east as the mouth of the St. Lawrence River. These are believed to have grown on unconsolidated sea-bottom in a substratum of limy clay, into which they have sunk. The various reefs are usually separated from one another, and the average size is about one-half mile across. They consist of a central core area surrounded by a layered area formed by material broken off from the core. They were built up of corals, sponges, crinoids, bryozoans, trilobites, cystoids and blastoids (relatives of the starfishes), and shells. Reefs of this type are common in other formations in the Midwest.

Alberta produces two-thirds of Canada's oil. Most of it comes from porous limestone of reef origin with typical Devonian fossils. The rock is composed of broken fragments of corals and like types which formed the detritus alongside the reef.

Fossils, Flood, and Fire

How much of the reef material lies in its original position and how much owes its position to disturbances, is difficult to say. Near Chicago the reefs are interbedded with shales, which indicates considerable strength of currents. From New York to Illinois the coralline beds of limestone have from 10-15% of clayey material that drifted in among the reefs. The question as to the location of the bottom of the reef is one that cannot be answered with any degree of certainty. It is difficult to prove any time relationship.

In western Texas and eastern New Mexico is a great reef formation, the Capitan, which is classified as Permian, but seems to be sunk into sediments that had accumulated before it was built up. It consists of 2,000 feet of dolomite (magnesium limestone), is about five miles wide, and is exposed along the flank of the Guadalupe Mountains for about 70 miles. Debris from the reef has rolled down the face and formed a slope of talus that becomes finer and finer the farther we look down the slope. Back of the reef is an accumulation of different nature, apparently formed in a lagoon. Here, then, we see three types of material in close proximity, but all related to the reef complex.

Could these reefs have grown before the Flood? Obviously they did, for their relation to other sediments indicates they were buried along with other materials by that great catastrophe. Generally the underlying material is such as may have accumulated in quiet seas. As to the time required, it must be admitted that we know so little about conditions in existence before the Flood that we are not in a position to judge the rate at which these reefs may have been built up. Anyone who insists that we must stretch time back to allow for a build-up at the same rate reefs are now growing is making an assumption that is unwarranted.

The fact that the reefs are classified all the way up to the Permian means nothing to the diluvialist, for

Life of the Ancient Seas

he understands that they all were contemporaneous anyway, and the supposed "age" is merely an indication of zonal distribution in relation to other beds that contain fossils by which their supposed time value is estimated. It is hard to cast off the uniformitarian "frame of reference," but if we can do so, the problem will be tremendously simplified.

Another striking feature of the Devonian rocks is the presence of armored fishes. Some of these were 20 feet long. The front part of the body was covered with heavy armor. They were bottom feeders, sucking up food from the mud.

The various animals we have described are not confined to the strata under discussion. Many of them, as in the case of the reef-builders, are found as high up as the Permian. A few, notably the crinoids, exist today. The distinction between the different "periods" is difficult to make. The rocks are similar in nature throughout, and only by fossil assemblages can the different formations be distinguished. There is little evidence of a change in the nature of the sedimentation. They all, as far up as the Mississippian at least, seem to have been formed under similar conditions. It is worthy of note that the Ordovician system was set up originally as a compromise between the views of Sedgwick and of Murchison; there may be serious question as to its validity as a separate system.

Another point should be noted. On the Colorado Plateau the Ordovician and Silurian rocks are practically absent, and in some cases the Devonian is not certain. This brings Devonian or Mississippian to rest directly on the Cambrian. In the Arbuckle Mountains of Oklahoma no distinction can be made between the Cambrian and the Ordovician. So it can readily be seen that exact classification of the lower Paleozoic rocks is not what one would expect if they were laid down over long periods of time.

Fossils, Flood, and Fire

As we review these marine fossils of the Paleozoic rocks, we are struck by the fact that so many of them are either attached to the sea-bottom or are bottom feeders. The fact that some of them are abundant in one formation more than in another is exactly what we would expect if local zonal groups had been buried.

One fact should be recognized, and that is that these different "periods," as the geologists call them, were not necessarily successive in time of deposition. If, as we assume in basic principles of the Flood theory, all these types of life were contemporaneous, then there is no way to decide the exact order in which widely separated formations are related chronologically to others. To illustrate, the "Permian" reef of the Guadulupe Mountains may have been growing at the same time as the "Silurian" reefs of the Midwest. And it may have been buried by the catastrophic action that buried the "Pennsylvanian" coal beds. In fact, types in it are similar to those usually found above the Pennsylvania strata and therefore classed as Permian.

The lower rocks of the Paleozoic systems appear to have been derived from several sources. On the east of the Appalachian Geosyncline was a continent, known as Appalachia, off the middle Atlantic coast, and Llanoria off the southern portion. From this highland much of the sedimentary material of the Appalachian region must have come. The sandstones in the Ordovician rocks and upwards seem to indicate considerable volume of washing. The geosyncline apparently filled rapidly, sometimes even with conglomerates, which are always formed in water in violent action. In the Devonian it is quite noticeable that the coarser sand grains are on the east side of the trough of the geosyncline and grade into finer and finer materials toward the west.

In addition to material from the east the Appalachian Geosyncline seems to have received sediments

Life of the Ancient Seas

from the Canadian Shield. Then in the far west the Cordilleran Geosyncline received materials from a highland farther west, known as Cascadia. No remnants of these highlands are preserved today except the Canadian Shield. Several smaller highlands, particularly in the Midwest, contributed locally to the Paleozoic rocks.

At the bottom of the Cambrian rocks a characteristic deposit of sand, conglomerate, and detrital material is to be found. How this was formed is hard to say, but the fact that in many places this has been classified as pre-Cambrian makes its reasonable to consider it as basic material left in the bottom of the seas at the time of their formation. It is quite impossible to explain how such material could have spread over the bottom of the seas after life had been in existence for thousands of years.

Some geologists have suggested that these sands may have been due to changing sea levels during "pre-Cambrian times," which would cause a migration of beaches. Looking at this idea in light of the Flood theory, we might find some value in it. Possibly the rising Flood waters may have acted on ancient beaches and distributed some of their materials. While this is only speculation, yet it may be worth studying.

A word of caution may be well here, lest the reader get the idea that because the rocks in different parts of the world may have the same names, such as Cambrian, Devonian, etc., their lithological composition or fossil content will always be the same, no matter where located. Lithological composition varies from place to place. It is not a dependable guide to the "age" of a rock. And as far as fossil content is concerned, there is no uniformity. While the rocks of Europe and eastern North America are fairly alike, yet they differ greatly from those of western North America. The classification is based on the presence of

Fossils, Flood, and Fire

a few "index fossils," as they are called. Often the selection of these index fossils is quite arbitrary, depending on the judgment of the expert who works on the specimens brought in from the field. It is impossible to prove that two rocks, because they contain the same index fossils, can be correlated as to "age." There is so much abstract theory and so much assumption in paleontological determination that the whole theory of the age of the rocks is an open question, although every geologist accepts it as gospel truth.

The fact is that about seven criteria are used in correlating strata from different regions. These are: the order of superposition, presence of fossils, lithological characters, stratigraphical continuity, unconformities, degree of metamorphism, and tectonic disturbances. These certainly are reasonable guides to the relative sequence of formations, *provided* that they have been produced under uniformitarian conditions. But it is easy to see that the theory of uniformity must be accepted to make these rules effective.

Coming back now to our discussion of the lower rocks. The upper part of the Cambrian and much of the rest of the series up to and including Mississippian consists largely of limestone, although as we progress upward more and more sandstone appears. This is quite what we would expect the rising Flood waters to produce.

When we study the Ordovician rocks of the Appalachian region we observe that what is known as the Cincinnati Arch served as a barrier to the washing of the sediments farther westward as they were brought in from the highlands on the east.

At the time the Ordovician strata were being laid down, disturbances showed at the north end of the region, producing folds and thrusts. They were accompanied by volcanic action, which continued into higher sediments. In the Pennsylvanian it is estimated that

Life of the Ancient Seas

scores of cubic miles of intrusions accompanied the foldings. These great tectonic movements resulted in much metamorphism. Outstanding examples of these activities may be seen in the Vermont marbles and slates. Close folding resulted in what is known as "Logan's line," running from Pennsylvania to the Gulf of St. Lawrence, marking the location where the intensely metamorphosed rocks were thrust over onto unmoved ones to the north and west.

The mountains in the northern end of the Appalachian region caused by the uplift of the Ordovician rocks show evidence of having been eroded, uplifted, and redeposited, thus smoothing off the contours of the region. Geologists explain this action as taking place during long periods of time, but we must realize that while the Flood was in progress many such movements might have occurred. For instance, in the Devonian we find what are called "second generation mountains," produced apparently by such a succession of events.

The Ordovician rocks of this region contain a large amount of black shales. These have been the cause of much discussion, and their origin is not well understood. Many geologists consider them to have been formed from ancient soils. This sounds reasonable, although in some places certain other factors may have contributed to their formation.

The Cambrian and Ordovician black shales appear to be similar to the black muds now being formed in depressions in the North Sea, Baltic, and other protected areas in the oceans, where fine sediments, mostly silts and clays, are known to be accumulating in basins and troughs where there is not sufficient current to disturb them. Possibly in ancient times the Midwest region of America had a variety of bodies of water with many shoals, islands, and channels. The fossils throughout this area are irregularly distributed, and

Fossils, Flood, and Fire

there is remarkably little material derived from the land. And so, the black shales may perhaps, in some areas at least, owe their origin to conditions of like nature to those now prevailing in some places.

We should not jump to the conclusion that the same conditions prevailed everywhere that lower Paleozoic rocks have been formed. Each region must be studied by itself and judged on its own merits. For example, in the Northwest the situation is extremely complex, and possibly not sufficiently clear to make it possible to interpret it correctly. The limestones of the lower Paleozoic are associated with volcanic matter and some conglomerates and coarse deposits that indicate violent water action. The western shores of the interior "seas" previously mentioned seem to have been located in Montana and Wyoming. Materials from the disturbed area on the west became mingled with the deposits of quieter waters to the east. During the early stages of Flood action, this region as far as the Pacific Ocean was taken in the grip of severe crustal movements which produced masses of debris and obscured the previously deposited sediments, making their interpretation difficult if not impossible. However, these disturbances were important in forming the topography of the western portion of the United States.

This brings up the question: How can these mountains be dated? Is it legitimate for the diluvialist to accept these chronological explanations? Of course if we take the position some do, that all rock strata are located erratically, with no order or system, then any chronological arrangement would be impossible. On the other hand, if we agree, as practically all students of geology now do, including the diluvialists, that there is some order and system in the rocks, then we might be able to arrange them in some kind of time sequence. Whether the age be in millions of years or whether it be in months, would depend on the background

Life of the Ancient Seas

"frame of reference," but the relative order would be about the same. If, for instance, the White Mountains of New Hampshire are dated as having come up in the Ordovician "period" or "Flood-stage," it would be because later rock strata have been found to lie upon or around them in such a manner as to make it possible to ascertain the order of their uplift in relation to other rocks. And so, when we speak from time to time of the origin of geological features in a certain place in the scale, it may be understood that we are following this method of dating them relatively without implying anything as to the actual length of time involved.

CHAPTER NINE

Burial of the Coal Forests

This phase of our study involves some of the most dramatic and the most astounding events of the whole story of the Flood. From what we have noted so far it seems evident that various actions were taking place simultaneously in different parts of the earth. For example, in the Appalachian Geosyncline there was some volcanic activity and the rise of some mountain areas, apparently due to crustal disturbances, even as early as the Ordovician deposition. On the other hand, in the Midwest it seems to have been quiet most of this time, although there is evidence of some overwash of material from the east. But when we come to the burial of coal, terrific activity is evident.

As far as marine life is concerned, the Devonian, Mississippian, and Pennsylvanian rocks are much alike. There are some different species in the various formations, but that is no more than we might find today in different areas. Such a situation does not give proof of an evolutionary sequence, but can readily be understood in terms of ecological zonation.

The rocks designated by the British geologists as Carboniferous were found to be separated in America

Fossils, Flood, and Fire

into two distinct divisions, which have been named Mississippian and Pennsylvanian. The principal reason for the separation is the presence of an eroded surface in the middle of the Carboniferous rocks.

One peculiar feature of the upper Mississippian and the Pennsylvanian is alternating series of rocks known as *cyclothems*. As an example, we might find calcite limestone and quartz sandstone alternating as many as ten times in a thousand feet. The Pennsylvanian of West Virginia has ninety cyclothems showing channeling, shifting stream beds, and other evidences of rapid changes, exactly what we might expect from violent Flood action.

The plant life involved in the formation of the Pennsylvanian coal beds begins to show up in the Devonian rocks. Here we find spore-bearing marsh plants. Geologists speak of "forests" of lycopods, scouring rushes, and simple ferns in the Devonian. Many of the specimens are badly broken up, indicating that they were washed about by strong currents. In many cases they are in such bad condition that it is impossible to identify the species.

Some of the tectonic disturbances at the time of the burial of the Devonian rocks are worthy of passing notice. In Europe two mountain ranges, the Variscan Mountains, were folded up. Later they became buried beneath other sediments. The same thing happened in America. A series of folds in Oklahoma and southern Colorado raised rather prominent mountains which are now practically buried beneath later sediments. In Oklahoma the remnants of these ranges persist in the Arbuckle and Ouachita Mountains.

In the Devonian from New York and Pennsylvania westward the deposits are almost entirely terrigenous, consisting of sands, silts, and clays, with little limestone. This indicates heavy land erosion. It would seem that after the Cambrian, Ordovician, and Silurian deposits

Burial of the Coal Forests

of this region were formed, as we have pointed out, in comparatively quiet waters, a change occurred, and vigorous action began.

Now we come to the greatest event of all, the formation of the coal beds. Plant materials were deposited over a vast area, both in America and in other countries. Pennsylvanian coal occurs in four principal fields in the United States: (1) the Appalachian, from Pennsylvania to Alabama, (2) Michigan, (3) Illinois, and (4) Mid-continent, from Iowa to Texas.

In the Appalachian region the streams rushing down from the eastern highlands deposited a succession of shales, sandstones, and other materials in which much vegetation was included, but little marine material. A vast system of deltas was formed, reaching the whole length of the Appalachian region. More than half the sediments in this area consist of shales and siltstones, with some irregular beds of sandstone. Only occasionally are there marine limestones. As the materials were spread westward, they were acted upon by vigorous currents and were spread out to form an aggraded plain. Thus a combination of torrents of land origin and great waves acted together to produce the deposits.

As far down as the Devonian rocks this delta formation can be seen. Geologists tell of streams bringing down red mud from the slopes of ancient Appalachia and spreading them over the lowlands thus forming great deltas. The sandstones are found largely in the eastern border of the region, grading into siltstones and shales, and eventually, farther west, into black shales and into limestones.

In the region of Colorado and New Mexico the action was noticeably rapid. Strata from Cambrian to Mississippian are only a few hundred feet thick, but when we come to the Pennsylvanian we find it to be as much as two miles thick. As a whole the Pennsylvanian "period" of deposition was one of intense de-

Fossils, Flood, and Fire

formation, with deep troughs, folds, and upwarps being produced over nearly all of the United States from New England to Colorado.

Not all of the sediments came from the east. The Oriskany sandstone of the Appalachians is a deposit of pure sand usually less than 50 feet thick, and is from two sources. A thin sheet not over ten feet thick, running from New York to Michigan, came from the Canadian Shield; a thicker sheet from Maine to Michigan, came from the east. In it are the first buried forests. Stumps as much as two feet in diameter occur in this layer. In it also are the lowest of the amphibians. How anyone can figure that normal geological action could lay down these sheets of sandstone is hard to imagine.

Popular theory attempts to explain the Pennsylvanian coal beds as having been derived from great bogs. But while it is obvious that the materials forming the coal did grow in lowlands, possibly adjacent to extensive waterways, this is no reason for concluding that they grew in bogs or swamps. You may go to Hawaii today and find a most luxuriant growth of tree ferns with no swampy areas whatsoever.

A great difficulty in the bog theory is the large number of alternating beds of shale and silt in between the coal beds. From 50 to 100 or more such successive beds may occur in some areas. This would require that the whole region, sometimes embracing hundreds of thousands of square miles, be lifted up just far enough to allow floods of water to sweep sediments over the bog and to bury it. This alternation would have to continue for millions of years. Such a finely balanced arrangement of land and water is extremely difficult to explain. Furthermore, in the great expanse of the coal beds, as in the Midwest, where was the high land that furnished the mud to bury the bogs?

Burial of the Coal Forests

If such alternating rising and falling of the land level had occurred, the whole region would have gone through a series of stages of sea-beaches repeatedly, both during submergence and emergence. Yet there is no evidence of this.

Again, if coal had been formed during the 50,000,000 years of the Pennsylvanian "period," as it is supposed to have been, it is a peculiar fact that the vegetation in the upper beds is identical with that in the lower beds. The identity is world-wide. There is no evidence of evolutionary progress during that time.

When all these things are considered, the Flood theory affords the best explanation for the formation of the coal beds. Furthermore, it affords an explanation for the fact that in some cases plant material has been badly macerated, while in others it is finely preserved, as if it had been washed into place quickly and buried before disintegration could have taken place.

Wave after wave dashing in upon the land would tear away the earth and soil and carry away great masses of trees and other vegetation to be buried in layers of mud. The shales and silts between the coal beds represent, possibly, the remains of the soil in which the forests grew.

Coal beds in Nova Scotia and New Brunswick, where the Pennsylvanian rocks are 13,000 feet thick, are described as having been deposited in great basins between mountains. The entire group is non-marine, that is, it consists of materials washed in from the land. Sections exposed near South Joggins (see page 62), at the head of the Bay of Fundy, contain stumps of trees buried in upright position. Original reports made on this region by Charles Lyell and William Dawson attempt to explain the formation as having been produced by a succession of forests that grew and died and were buried in mud which later hardened to form shale. Beds of fire-clay that sometimes lie beneath the coal

Fossils, Flood, and Fire

layers are interpreted as the ancient soil in which the trees grew.

This interpretation is open to serious question. It is hard to see how some of these clay layers only a few inches thick could have supported a forest of trees of sufficient volume to produce the beds of coal just above them. In fact, accumulating evidence is opposed to the idea that the underclays—that is, the fire-clays— were formed beneath the coal. They might have been what they are now when they were deposited.

A further problem is seen in the fact that sometimes upright trees are found extending up through several layers of rock and sometimes through several layers of coal. How these trees could have been preserved for the length of time necessary to accumulate mud and plant material around them under any normal conditions is difficult to imagine.

Many other questions might be raised, which throw great uncertainty upon the popular interpretation. Even a casual glance at the exposed cliffs along the arm of the Bay of Fundy where these beds are exposed gives a strong impression that they were produced by something entirely out of the ordinary. And simply because most of the stumps stand more or less upright is no reason to believe that they are now standing where they grew.

In the 1966 Annual of the Creation Research Society is an article by N. A. Rupke of the State University of Groningen, Netherlands. In this article he describes the problem of "polystrate fossils," that is, fossils that extend up through many layers. He says:

> Curiously a paper on polystrate fossils appears to be 'black swan' in geological literature. Antecedent to this synopsis a systematic discussion of the relevant phenomena was never published.

Many examples are given of tree trunks standing upright as much as 90° from the horizontal. Some-

Burial of the Coal Forests

times they are even found to be upside-down, which, of course, makes normal burial impossible. As to the question of whether the trees grew in place or not, note the following statement that he makes:

> It must be remembered that a drifted tree-stem will often float upright, for its center of gravity is situated at the lower end of that stem, which constitutes the heavier one. In this context Fairholme (1837) wrote: "The stem of a heavy tree, especially if it be long, and have consequently a great disproportion between the weight of its two extremities, would naturally sink in a fluid, and perhaps still more, in a semi-fluid, with its root downward. . . ." This actually has been observed along the banks of the great rivers; Arber (1912) writes: "In the deltas of large rivers today, the bases of trees of large size, sometimes with fragments of roots still attached, may be deposited in a more or less upright position, though they have been transported for a considerable distance."
>
> Moreover, . . . petrified tree-stems more often than not display a thickened trunk-base. . . . As a result they tended to float upright and were deposited in that position as soon as the current quieted.

From these statements we can see that there is a question as to the validity of the argument that if these stumps had been washed into place they would have been tumbled about in all directions.

The forests that were destroyed to form the coal beds were not at all like our modern forests. On the contrary, they were largely of spore-bearing trees which were related to modern club-mosses, ground pines, and scouring rushes. The most striking were the scale-trees, so-called because their leaves on dropping left scalelike scars on the trunks and limbs. These trees grew four to six feet in diameter and over a hundred feet high.

Lepidodendron (scale-tree) branched repeatedly, and each branch bore a bunch of slender leaves, which resembled huge pine needles. Some of them were six to eight inches long and half an inch wide. Spore cases

Reconstruction of a Pennsylvanian Forest. Drawing by Elden James.

Burial of the Coal Forests

resembling cones grew on the tips of the branches. The trunk was pithy at the center, with wood on the outside, and it had a thick bark.

Sigillaria had a thick trunk with few branches. This tree was about the size of *Lepidodendron*, and was thickly covered with long, narrow leaves.

About a hundred species of these trees have been found. Many of them were large; others were small.

The root systems were peculiar, the main roots extending horizontally with few branches. Many structures, known as *stigmaria*, sprang from the sides of the main roots, and radiated several inches. These are a great unsolved puzzle. There is no certain proof that they are really roots or rhizomes. Their exact nature and function are unknown. Their presence in the clay and other materials beneath the stumps has been taken as evidence that the trees were buried *in situ*, that is, where they grew. However, it may be pointed out that this same condition might prevail had the stumps been washed into place with their roots and rootlets still attached, provided there was sufficient earth clinging to them to preserve them. The fact that the stigmaria not infrequently occur in the coal beds themselves as well as in the clay lends credence to this interpretation. Violent washing, bringing with it a large amount of earth, might create a false impression of *in situ* burial.

Giant scouring rushes grew thickly in portions of the lowlands. They possessed the same jointed stems as modern scouring rushes. Some of them were a foot in diameter and thirty feet high.

Another group, the *Cordaites*, somewhat resembled pines, but bore leaves instead of needles. Some of the leaves were five to six feet long. Many of these trees were as much as 100 feet high. The branches were all borne near the top of the trunk, and as they spread out widely they gave a pleasing appearance to the trees. The wood was similar to that of modern pines, and

Fossils, Flood, and Fire

was apparently an important constituent of coal.

Members of the *Equisetaceae*, to which the scouring rushes belong, had a great deal of silicon in their stems, and this had a tendency to make the Pennsylvanian coal hard. When it burns, it leaves considerable "clinker" due to the incombustible mineral matter in the trunks of these trees. The hardness of the Pennsylvanian coal is partly due, also, to the fact that it was buried deeper than other coals, and has been subjected to greater pressure.

Ferns were common, and grew as much as fifty feet high. Among them was one group unknown today, the seed ferns. They bore seeds on the ends of the branches. They were more abundant than the true ferns.

Some of the animal life of the coal forests is worthy of mention. This supposed "period" has often been called the "age of amphibians." More than 75 different kinds have been described, some of them attaining a length of ten feet or more. One form found in Kansas is estimated to have weighed more than 400 pounds. Some appear to have lived entirely in water.

Naturally only a few reptiles have been found. The diluvialist finds in this a reasonable correlation with the Flood theory, as they usually prefer higher and drier ground than amphibians. Some are known, but they differ so much from modern reptiles that we would hardly recognize them.

Insects were common. Over 300 species of cockroaches have been described from North American rocks alone. Some of them were four inches long and had a wing spread of nearly two feet.

Most of the fishes found in the Pennsylvanian rocks are fresh-water rather than marine. They had shiny enameled scales like modern sturgeons.

Shellfish of various types were abundant. Other marine invertebrates such as starfishes and their relatives form some of the most abundant fossil deposits.

Upper: Fossil Amphibians. Lower: Fossil Insects. Drawings by Elden James.

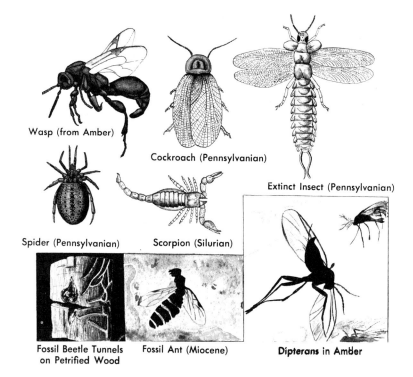

Fossils, Flood, and Fire

This indicates that the sea waters were involved in forming the Pennsylvanian deposits.

One problem that has given diluvialists concern is the presence of marine protozoans known as *fusulinids*. These look like grains of wheat. A closely related genus, *Schwagerina,* is also widely distributed. They are generally arranged in zones, and often these zones are used to correlate the beds as to relative "age." How could such a zonation occur in the ocean and be reflected in the coal beds, if they were overwhelmed by great waves?

It might be well to note that in the same beds there is a striking alternation of sandstone, shale, and limestone. This could be interpreted as due to alternation of wave action from different directions. If so, why could not the "zonation" of the fusulinids be due to the same cause? Instead of representing successive evolutionary stages for these creatures, why not think of these successions as representing deposits by waters coming from different directions. More details cannot be given here, but we suggest that such an approach might reveal hitherto unsuspected relationships.

About the time when the Pennsylvanian sediments in the Appalachian trough were nearing completion, great changes apparently took place in the land mass from which they were coming. Just what happened, of course we can never know, but the changes were of such a nature as to bring largely to a close the transportation of materials into the trough, although possibly not entirely. There is evidence that tectonic movements caused the ancient land mass to sink; and in so doing, to be shoved strongly against the materials that had been deposited in the geosyncline. This evidence lies in the fact that the rocks along much of the Atlantic coast are crystalline, belonging to the basement complex type such as are common in the Canadian Shield. Emerging from beneath the ocean,

A group of fossil Protozoa known as Fusulinids, magnified about 20 times. Photo by Ernest S. Booth.

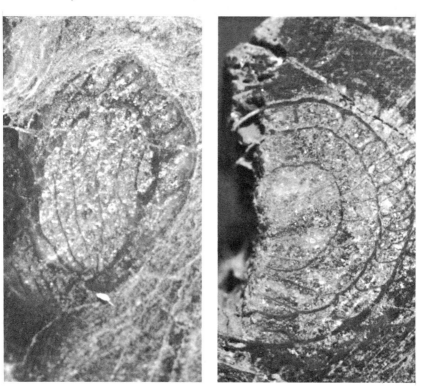

Two Fusulinids magnified about 50 times. Photos by Ernest S. Booth.

Fossils, Flood, and Fire

these rocks form a low, flat coastal plain that rises gradually to the foot of the Blue Ridge Mountains, then rises abruptly to form this ridge that seems to have been thrust vigorously over and against the sedimentary rocks to the west of it.

This thrusting force from the east caused great folds in the Appalachian sediments, forming a series of intensely folded lines of mountains. After this action, few sediments were deposited in this region, and heavy erosion wore away the tops of many of the ridges. Probably thousands of feet of material were denuded and carried away.

The question may arise as to how this erosion could take place if the region was uplifted above the ocean. But, thinking in terms of the Flood, we must not think that after the forty days of rain described in Genesis the skies would clear and the rest of the voyage of the Ark would be over quiet seas. On the contrary, terrific precipitation would cause severe flooding everywhere, even if there was no more washing by sea water. The sediments being soft, would wash away rapidly.

CHAPTER TEN

Permian Puzzles

When Murchison and Sedgwick, in the 1830s, had established the classification of the British rocks up to and including the Carboniferous, they became interested in the sandstones lying immediately above the coal. Investigation on the continent had shown that practically the same sequence of fossils occurred there as in Britain; and in the province of Perm, Russia, they found the red beds above the coal to be rich in fossils. It looked as if these sands had been washed into a great basin, where both marine and land fossils were mingled.

As James Hall and his associates did their work about 1840 on the rocks of the Appalachian region, they found few areas where the Permian could be recognized. If these sands were deposited, they had largely been swept away by erosion. From the size of the folds of this region it is apparent that they must have stood several thousand feet high as they were shoved up after the deposition of the beds.

In the Midwest, Permian sands are thin, and are interbedded with shales and thin limestones. For a long time it looked as if Permian rocks were not of much

Fossils, Flood, and Fire

consequence in America, until 1900, when a thick and richly fossiliferous region was discovered in west Texas and New Mexico. This is commonly spoken of as the Texas Permian Basin. Here the rocks classified as Permian are as much as 14,000 feet thick. The most striking feature is the great reef forming the Guadalupe Mountains. Materials washed off the reef had apparently been accumulating for a considerable length of time, and were now mingled with sediments brought in from the land mass known as Llanoria, lying off the present Gulf of Mexico and southern Atlantic.

Geologists speak of the Permian as a "period" of violent and rapid action and of extremely fast geographical changes. In a number of places in the world great diastrophic (earth-changing) movements took place, upheaving mountain ranges, and shifting the course of the waters that were acting on the earth. It is difficult to gain a clear understanding of what was really going on, as all geological literature is written in terms of uniformitarianism. But if we translate "period" or "ages" into *stages,* we may be able to form some kind of a picture of events transpiring.

One notable feature of the Permian rocks is the great reduction of marine fossils and the increase of land life. But many of the rocks are poor in fossils. In many places the rocks already laid down were being thrown up and folded and thrust over one another, and the waters were carrying away much of the material and redepositing it in low areas. Under such conditions what plants and animals were present in the newly upheaved rocks would not have time to thoroughly fossilize, and would readily be destroyed. This fact may account for the comparatively few fossils in the Permian rocks.

One of the most puzzling phenomena of the Permian is the presence in certain regions of *tillites,* deposits of cobblestones, gravel and other materials that re-

Permian Puzzles

semble glacial moraines. On this evidence it has been assumed that glaciation took place in the Permian "period." Probably the most extensive such areas are the so-called glaciated regions of South Africa. These have been studied for many years, and are still one of the greatest of geological conundrums. Writers on geology admit that there is no acceptable explanation for supposed Permian glaciation, although they can find no other explanation that satisfies them. Among the problems that face the geologists is the theory that ice moved from near the equator to the cooler climate of the south. A review of the situation brings out other problems as well.

One of the earliest discussions on the supposed glaciation in South Africa was held in the Geological Society of South Africa in 1897. Scattered boulders mixed with a clayey matrix, scratched and striated rocks, were given as proofs. Some members disputed this conclusion, declaring that other forces could have produced the results. In 1921 Alex. du Toit, in an article in a local geological journal, stated that the problem was as obscure, elusive, and complex as ever. And then, eighteen years later, in the *Geology of South Africa*, he spoke of the exceptional difficulties and the extremely restricted evidence on which the glacial theory rested, calling it a long-time puzzle.

Other writers have described the stones in the South African breccias as resembling gravel fans arranged by water, rather than true glacial debris. The most typical of these, the Dwyka tillite, resembles volcanic tuff. It consists of pebbles and boulders derived from the underlying rock, and is sometimes scratched and polished. But in places the deposits pass downward without break into deep-water shales, or grade into beds of limestone. The intergrading seems to make it impossible to interpret the loose material as true glacial debris. Also, a map of the Dwyka tillite shows no relation that would coincide with the direction of the striae, and no evidence that it is a true morainal deposit.

Many other criticisms of the glacial theory of the tillites were given in discussions and papers published around the beginnings of the century. In the Congo Basin similar pebbles have been found. But it is hard to imagine glaciation in that tropical region.

Deposits similar to the tillites of South Africa have been found in South America. The Falkland Islands have 10,000 feet precisely like the Dwyka, and a great tillite body lies in Argentina and Paraguay. Regarding these du Toit says that the evidence is "disappointingly meager." Tillites are also found in Madagascar and in India, but again the evidence is insufficient to be of much value. Polished, faceted, and striated blocks suggest a fluvio-glacial agency. The Salt Range beds of India resemble a marine deposit. They are not morainal. Tropical mollusks occur in the beds, and they are overlaid by true marine sandstone. Most of the conditions prevailing in India hold good for Australia. It is apparent that the ancient "Gondwanaland," which extended nearly around the world, was overwhelmed by a great catastrophe which swept over all parts of it at the same time.

It has not been easy for advocates of Permian glaciation to prove that their conclusions are valid. Even for men who believe in long geological time, the evidence has been confused and conflicting. At a meeting of the British Association in 1893 it was declared emphatically that the kinds of rock found in the Permian breccias of England had their exact counterpart *in situ* in the immediate vicinity, and that there was no certain evidence of glacial action.

While Murchison suggested ice action for some of the beds of breccia, his claim was disputed, and it was pointed out that they differed very definitely from

Permian Puzzles

ordinary glacial deposits. They are interbedded with finely laminated shale. There is no sorting of materials, no evidence of glacial polish or striation in the British breccias. The alternating clay and shale beds, it was declared, indicate a change from quiet water to violent floods. The Permian breccias, it was stated before a meeting of the London Geological Society in 1894, show all the characteristics of great gravel fans. The Indian and Australian "glacial" beds are such as would be developed by dropping fine sediments in quiet water, mixed with transported fragments.

While many of these references are from papers or speeches made many years ago, it may be significant that the men who were then struggling with the problem candidly faced the facts. Since that time new generations of students have accepted the glacial hypothesis without realizing the inconsistencies it offers. Doubtless the opinions of these pioneers as they dealt with the problem may be of more value than the ideas expressed by current writers. And it is particularly interesting that the greatest student of South African geology, du Toit, after more than forty years of study, still recognizes the evidence for glaciation as "disappointingly meager."

Most recent textbooks of geology take Permian glaciation for granted, and the student reading them would never guess the problems that this theory involves. However, an occasional discussion does admit the difficulties. One such text speaks of the fact that such an ice age is very difficult to explain because the evidence is centered in what is now tropical or semi-tropical regions. Furthermore, most of the movement appeared to be from the equatorial regions. This is especially puzzling in India, where the scratches on the rocks seem to indicate a northward movement of the ice, whereas it would be expected to be just the opposite. Much speculation has been indulged in, at-

Fossils, Flood, and Fire

tempting to iron out the inconsistencies, but they remain to plague the geologists in spite of all their efforts to unravel them.

More recent studies on tillites have brought to light new ideas that would appear to make the explanation in terms of glaciation obsolete. Ernest J. Opik, in his book *The Earth and its Atmosphere* (1957), says that this is one of the greatest geological puzzles. R. F. Flint, in his *Glacial and Pleistocene Geology* (1957), says that striations can be made by floating or flowing mass. Other agencies, then, besides glaciers, can produce *till*. At a meeting of the Geological Society of America in 1959 it was declared that most ancient tillites and glacial "periods" must be regarded with suspicion.

Of course it must be recognized that there are other apparently valid evidences for glaciation besides till and tillites, and these will be more fully considered in the discussion of the Pleistocene rocks. But for the present we leave the subject of Permian glaciation as being largely a theoretical consideration that is far from being based on sound evidence.

Another problem of the Permian rocks, and of some other systems, is the presence of large deposits of salts. These are supposed to represent long ages of slow evaporation. It is a peculiar fact, though, that there are on earth today no seas where such salt deposition is in progress. It may be that we need to look for some other cause of the salt deposits. The suggestion has been made that profound changes in the electrical constitution of the atmosphere at the time of the Flood might have brought about changes in the composition of the ocean waters. No definite proofs have yet been presented, but the subject is one worthy of investigation. It may well be that there were influences at work of which we have no knowledge whatsoever, and that we should be cautious before we attempt to explain everything in the past by what we can see in operation today.

CHAPTER ELEVEN

Mystery of the Red Beds

As we progress in our studies, up through the Permian and into the Mesozoic rocks, we find a sudden change in the fossils, although not much in the nature of the rocks themselves. The Triassic and Jurassic rocks are often spoken of as the "red beds," because of the abundance of red sandstone which they contain. However, the red sandstone began as far down as the Devonian, which was originally called the Old Red Sandstone.

The Triassic, Jurassic, and Cretaceous rocks are grouped as Mesozoic (middle-life) rocks, because they apparently constitute a distinct change from the Paleozoic, yet are markedly different from the modern types, which are called Cenozoic (recent-life).

The name Triassic was given to the lowest of the three because in the continent of Europe it had three divisions. This is not always true in other parts of the world, in fact, is generally not true at all. Jurassic was named from its abundance in the Jura Mountains of France; and the name Cretaceous comes from the Latin *creta*, chalk, and applied to these rocks because they embraced the famous chalk cliffs of England. The distinction between these three groups is not always clear, and many features of which we shall speak may be applied to any of them.

Fossils, Flood, and Fire

As a whole, however, Triassic and closely related rocks are widely distributed over the world, and are much alike in general character. In Brazil Triassic lava flows cover 300,000 square miles. Extensive flows are found in South America. A peculiar phenomenon of Triassic rocks is the imprints of raindrops. This would indicate that there was time enough between "showers" to allow the mud to harden and preserve the imprints.

The only Triassic rocks in the eastern United States are on the eastern edge of the northern Appalachians, from North Carolina to Nova Scotia. They consist of a mixture of red sandstones, siltstones, and conglomerates, which were evidently washed down from the uplands by torrential streams, and settled in troughs formed by tectonic movements in the rising Appalachians. The sediments are poorly sorted, irregularly bedded, and the sandstone grades into siltstone and clay. Conglomerates are thickest on the eastern side, indicating that the wash was from that direction. Mixed with the sediments are extensive lava flows, such as those forming the Palisades on the Hudson.

Another feature is the presence of cyclothems, such as we noticed in the Pennsylvanian. Hundreds of successions of land-derived materials occur in some of the rocks. Since these are of land origin they cannot be ascribed to rising and sinking of the land, and so they present a real puzzle to the geologists. How any such conditions could prevail on land as to produce these cyclothems, is a mystery. But from the viewpoint of diluvialism the solution is much simpler, in fact is almost obvious—merely a matter of successive washing or flooding.

The Triassic Newark beds in the vicinity of New Jersey and New York lie on pre-Cambrian gneisses and schists, with none of the supposed "ages" in between. This is explained as being due to the sediments having been eroded as the Appalachian uplift occurred, leaving

The remarkable red beds of the Southwest. Upper: Rainbow Bridge, Utah, largest natural bridge in the world. Lower: Massive sandstone cliffs in Rainbow Bridge Canyon on Lake Powell, near Rainbow Bridge, Utah. Photos by Ernest S. Booth.

Fossils, Flood, and Fire

the exposed pre-Cambrian rocks on which the "Triassic age" materials were laid down. From the angle of long "ages" where did the eroded materials go, and where did the new deposits come from? Again, the diluvialist has a viewpoint that makes an easier explanation than the long-ages geologists can suggest.

Triassic beds are almost entirely non-marine the world over, and one of the most striking display of these rocks occurs in the great Colorado Plateau and nearby areas. Not only are they interesting, but they present one of the greatest mysteries of the whole geological column.

If we go northward from the Grand Canyon, either on the east or the west side of Utah, we find red beds spread out everywhere, with beautifully carved cliffs, monuments, and canyons. At Moab, a small city in the eastern side of the state, beautiful red and brown cliffs rise 2000 feet above the Colorado River. These are classified as Triassic and Jurassic. Well logs show from 7,000 to 10,000 feet of sediments in this area. A few miles north of Moab, on the north side of the desert plateau, are 2000 feet of Cretaceous cliffs that overlie the red beds. All in all, the eye can see about two miles of sediments, mostly red sandstones, but with some limestone, shale, and conglomerate spread out between the sandstone layers.

The mystery appears when we begin to examine the lateral distribution of these beds. From the northern part of Arizona to southern Wyoming, between 400 and 500 miles, this great Colorado Plateau, as it is called, appears to be one great sedimentary basin of approximately 200,000 square miles. Estimates of the amount of material deposited here before erosion washed any of it away run as high as a million cubic miles, and in some cases even more. What a movement of rock-forming sand and mud must have taken place to produce these tremendous deposits!

Unusual geological formations are carved from the red beds. These two photos show Delicate Arch in Arches National Monument, eastern Utah, one of the most remarkable geological wonders in the world. Photos by Ernest S. Booth.

Fossils, Flood, and Fire

Where did it all come from? And by what means did it get here? These are simple questions, but the answers involve some of the most profound mysteries of the past. To understand the problem, we must go into a few technicalities, for that is the only way to get a picture of what has happened in the past. The rocks tell their own story, so let us ask them what they can reveal to us.

In the vicinity of Moab several formations are displayed on the cliffs west of the highway. Their names are:

TRIASSIC
    GLEN CANYON GROUP
        NAVAJO
        KAYENTA
        WINGATE
    CHINLE
    SHINARUMP
    MOENKOPI

(NOTE: Geological strata are always diagrammed as they lie, with the lowest formations at the bottom.)

Just below these are the Paleozoic rocks, but they do not come to the surface; they do, nevertheless, show up in the well logs.

Now let us examine the Triassic rocks in some detail.

The Moenkopi formation consists of up to 500 feet of red and brown mudstones, shales, or muddy sandstones, and it weathers to form brilliant slopes. It contains few fossils, but what are found lead to the conclusion that the beds are of continental origin, not marine.

The Moenkopi underwent structural warping before the overlying sediments were laid down upon it. Yet no deep canyons appear anywhere. The next higher formation, the Shinarump (pronounced: SHIN-ah-roomp) was dropped into shallow hollows on top of

These red beds found in Canyonlands National Park, southern Utah, have been greatly eroded by wind and water. Both photos show "The Needles," in Canyonlands. Photos by Ernest S. Booth.

Upper: Rainbow Forest in the Petrified Forest National Monument, Arizona, shows large numbers of petrified logs of a number of species of trees. Lower: A single petrified stump extends above ground on Amethyst Mountain, Yellowstone National Park, Wyoming. Photos by Ernest S. Booth.

Typical Mesozoic strata in the red beds at Capitol Reef National Monument, central Utah. Upper photo by Harold W. Clark; lower photo by Ernest S. Booth.

Fossils, Flood, and Fire

the Moenkopi. These hollows are not over forty feet in depth, and in many places are less. We are puzzled as to how geologists can believe that millions of years are involved in these processes without leaving deep canyons instead of shallow hollows.

The Shinarump is a peculiar type of formation. In places it is a hard, resistant sandstone containing many small pebbles—in other words, it is a sort of conglomerate. The pebbles are well rounded, showing that they have been washed for great distances. They are usually less than two inches in diameter, and are composed of quartz, quartzite, chalcedony, and flint, of various colors. Where did these pebbles come from? There are no rocks of their composition in the underlying Moenkopi or any other nearby formation. Yet, according to the popular theory of the "ages," Moenkopi rocks should have formed the landscape from which streams would wash out the materials to form the Shinarump deposits. And the problem is not a local one, for what we have said applies to the whole Colorado Plateau. This peculiar Shinarump formation is spread out over more than a hundred thousand square miles.

In many places the Shinarump contains sandstone and shale as well as conglomerate. At the top it grades into the Chinle, so that the distinction between the two is hard to make. The Chinle consists of sandstones, shales, mudstones, and conglomerates. These various types integrade. They show considerable irregularity in local bedding, as if strong streams and whirling waters had dumped their loads into bodies of water. This "delta" bedding is true also of the Moenkopi.

Fossil wood occurs in "log jams," which is another indication of flood-plain or delta conditions with rapidly running water.

The Chinle has many small dinosaurs. One of its most interesting features is the Petrified Forest of Arizona, which has the most wonderful display of petri-

Upper: Petrified stump of redwood at Amethyst Mountain, Yellowstone National Park, Wyoming. Lower: A piece of petrified redwood lies on top of stump at Amethyst Mountain, Yellowstone National Park. Photos by Ernest S. Booth.

fied wood in the world. Some of the trees were nearly 200 feet high. They were conifers, something like pines, as were nearly all Triassic forests. There is no evidence that the trees actually grew where they now lie. Apparently they were rafted into place by great streams of water.

Above the Chinle lies the Glen Canyon group, consisting of three formations of cross-bedded sandstones. Their boundaries are indefinite, and they often intergrade. Wherever they are exposed, they form vertical cliffs totaling from 700 to 1000 feet. The bedding is irregular. In some places cuts of fifteen to twenty feet have been made before the next layers were deposited, but beyond this slight irregularity no signs of deep erosion are seen.

On the east side of the highway at Moab another group of formations lie exposed in the Arches National Monument. They lie above the Triassic as follows:

JURASSIC
    MORRISON
    SUMMERVILLE
    ENTRADA
    CARMEL
TRIASSIC

Due to a fault, the Carmel, which normally lies on top of the Navajo, the top layer on the west side, is brought down to the level of the valley; and so as we go into the Monument to the east, we can follow the Jurassic series upward to the Cretaceous.

The Carmel is from 125 to 150 feet thick, and consists of pink to red or brown sandstones and mudstones, irregularly bedded. On top of it lie 250 feet or more of Entrada, a massive reddish-brown sandstone. The Summerville is less than 50 feet thick, and varies in composition. In one area it contains great masses of agatized or opalized material.

Mystery of the Red Beds

One of the outstanding formations is the Morrison, which crops out a few miles to the north and east. This has many variations—sandstone, conglomerate, etc.—similar to the Shinarump, with shales, limestones of various colors, mudstones, and quartzites. The Morrison has been traced for over 100,000 square miles, and is seldom more than 400 feet thick. It shows up as far east as Oklahoma and North Dakota. Many of the giant dinosaurs are found in this formation. Present also are turtles, crocodilians, and other reptiles, similar to the ones found in the Triassic of Europe. Geologists say that the Morrison appears to have been laid down by rivers sweeping over extensive flood plains.

Above the Morrison lies the Dakota, a Cretaceous formation similar to the Morrison in superficial appearance, but made up of sandstone and clay largely.

We have gone into considerable detail in describing the rocks of the Colorado Plateau in order that the reader may have a clear picture of the problem. Now what conclusion can be drawn from these facts? Let us try to evaluate the data.

Of course the first answer that one would propose would be that the material forming these beds was washed down from nearby highlands and deposited in an ancient sea. But where were these highlands? Not the Rockies, nor the Wasatch, nor the Uintas, nor the La Sals, for all of these great mountain regions surrounding the Plateau were pushed up after most of the sediments of the region were laid down. They are uplifts and intrusives that have been forced up from deep down, and in coming up have often warped and twisted and distorted the overlying sediments. Only one local area seems to have contributed much to the rocks of the basin, and that is the Uncompaghre Plateau in southwestern Colorado, a mass of exposed granite rocks. But what it may have given could have

Fossils, Flood, and Fire

involved only the sediments up to and including the Permian. By the time the red beds began to be formed, they overlapped the Uncompaghre region and buried it. Today remnants of the red sands still may be seen on top of the crystalline rocks in places.

Geologists are able, by examining the thickness and texture of the rocks, to tell from which direction they have been derived. In the case of the rocks of the Colorado Plateau, most of the sandstones are believed to have come from the west and southwest. In some areas the formations are much thicker on the west than on the east; they thin out to the eastward. Also the materials become finer and finer as we go eastward. For instance, the Shinarump formation has the largest conglomerates along the Mogollon River in northern Arizona, and the size of the pebbles decreases toward the northeast. In Monument Valley it grades into sandstone.

It is true that recent reports attempt to describe "seas" or lakes scattered over the Plateau, into which sediments were washed from different directions. But this evidence is only what might be expected with local washing making deposits here and there. The local back and forth action of Flood waters should not be allowed to confuse the general over-all picture.

Studies farther west fail to reveal the source of the red sands. Central Nevada mountains contain as much as 15,000 feet of lower marine sediments, but there is nothing that could have supplied the red sands. The conclusion seems inevitable that an ancient continental mass in the vicinity of California or the western Pacific must have been the source.

A striking fact that supports this conclusion is the remarkable evenness of depositions, with little erosion such as goes on today. How beds of sandstone, mudstone, and shale could have been exposed to the at-

Mystery of the Red Beds

mospheric elements for millions of years and yet show no canyons or deep gorges such as recent times have produced, is indeed a mystery. Catastrophic flood action seems to be the most satisfactory answer.

Returning to the area itself, and examining it internally, we might ask if there might be some local highlands inside the boundaries that could have furnished the sands. The writer has gone from north to south and from east to west across the area, looking for any possible sources. Thousands of pages of reports have been scanned, but without avail. Nowhere has it been possible to find evidence of local sources.

Had the Chinle, for example, been derived from the Moenkopi landscape, the latter would have had to be elevated so as to have material washed from it to supply the Chinle beds. There is no sign that this has happened. Everywhere the Moenkopi lies below the Chinle. How could it have produced the Chinle beds unless it had risen higher than the Chinle?

As one studies the cliffs, even though they expose a supposed 70,000,000 years of deposit, nowhere do they show land masses, mountain ranges, badlands, river-carved canyons, or beach lines such as might be expected in any normal course of events. Look at what has happened in that region in recent times; even if we allow the time lapse accepted by popular geological theory, not over a million or so years would be involved. But geological forces that could spread pebbles over hundreds of miles must have been much more violent than any recent actions.

One more peculiar fact must be pointed out. Several of these successive formations are very similar, in fact, so nearly identical in composition and appearance that it is difficult to identify them unless we can follow through their sequence. In one locality the great cliffs are made of Wingate, in another of Chinle, in another of Entrada. There is an alternation of massive sand-

Fossils, Flood, and Fire

stone repeatedly. The same is true of some other rock materials. How this could have happened over 70,-000,000 years of time is hard to understand.

After all the evidence has been assembled, several obvious conclusions seem to be justified: (1) the sediments have come from a great distance, not from local sources; (2) they were brought in by great sweeps of water, for no ordinary rivers could spread masses of pebbles over such wide areas; and (3) they were laid down one after the other in rapid succession, with no long periods of erosion between. On what other basis can we reasonably explain the evenness of the contours between the formations, the irregular bedding within them, and the alternate occurrence of the massive sandstones and conglomerates? The Genesis Flood seems to be the only reasonable answer to the mystery of the "Red Beds."

CHAPTER TWELVE

The Violent Climax

The Mesozoic strata, although in many places sparsely supplied with animal fossils, do contain many plant species. Among them are some of the same types already noted in the Pennsylvanian, such as ferns and scouring rushes. But the bulk of the material consists of upland plants, among which may be found such forms as conifers and cycads. By the time the upper levels of the Cretaceous are reached, we find many flowering plants, which must have dominated the landscape. We might mention figs, sassafrass, poplar, and magnolia, then above them were hardwoods as in modern forests—birch, beech, maple, oak, walnut, sweet gum, laurel, hazelnut, and holly. Among the conifers was the Sequoia. There were grasses, palms, lilies, iris, and orchids, too. At least 30,000 fossil plants are known from the Cretaceous. It is clear that a very different ecological zone must have separated these plants from those we saw in the Pennsylvanian rocks.

Not only are land plants abundant, but insects as well. For instance, in the Jurassic rocks about a thousand kinds of insects have been found, all similar to modern types. They include caddis-flies, dragonflies, beetles, grasshoppers, cockroaches, termites, and moths.

Fossils, Flood, and Fire

Jurassic rocks are famous for dinosaur remains. The Morrison formation, extending over a great expanse of territory in the Rocky Mountain region, is particularly rich in their remains. The large number of land fossils which it contains, the poor bedding, the poor sorting, and the presence of wind-blown loess and sand dunes show clearly that it was produced from continental sources. Spread out over about 100,000 square miles, and usually only about 400 or 500 feet thick at the most, it indicates clearly that it must have washed into place by violent streams of water, and not by any normal type of action. The rocks are sandstones, shales, siltstones, conglomerates, and limestones, and the bedding is hard to follow. No marine fossils are known.

Here we have a situation that may raise a question for anyone trying to explain the formation of these rocks by Flood action. How could there be sufficient water to spread out such beds unless ocean waters swept over the land? Now it is evident that at the time the Morrison formation was being laid down, ocean waters were being swept over the lands in some places, for in the Jurassic is found a rich assemblage of marine fossils. Among them are corals, oysters, lobsters, crinoids, sea urchins, and sponges. But we do not need to feel that we must invoke ocean waters in order to explain the engulfment of all the great land animals. It must be recognized by anyone who stops to think the matter through carefully, that extensive "seas" may well have existed at different levels, and that waters from these may have buried the land animals in some instances. Geologists who see nothing but long ages of time would of course interpret these beds as having been laid down in a period when the land was above the sea and torrential rains swept down mud to bury the dinosaurs who lived around the shores of these seas. But it is still perfectly logical to assume that they may have been buried by waters from inland seas or

The Violent Climax

lakes. In fact, it would be difficult to attempt to separate the ancient waters into ocean and lake waters, into marine and land sources. Our modern classification might not fit the situation at all.

This principle must be kept in mind when we notice from time to time that different species of water life appear at different levels. Their presence may be understood in the light of the principle of zonation. For instance, clams and other water life may occur at different levels in their particular zones. This idea of zonation may explain their occurrence in different formations: it is not necessary to assume that the ocean waters bearing "marine" types were always responsible for the occurrence of what we may call "sea-life." The reason is that we really do not know whether the original waters of the earth were salty or fresh.

The most famous of the dinosaur beds is located near Vernal, Utah, where a bed of sandstone 150 feet long and 50 feet high is packed with bones. This cliff was once much higher, before it was reduced by the digging operations of the Carnegie Museum and others. The government has now erected a building against the cliff, where visitors may stand on a balcony and view the excavations that have exposed the dinosaur bones. The bones do not give evidence of natural burial, for they are badly disarticulated. They show every evidence of having been washed into place by violent currents of water.

> The reptiles found as fossils in the rocks range from the size of a small dog to monsters over 80 feet in length. Their habits of life were equally varied; for they embraced some which ate vegetables only, some which ate other animals only, and some which apparently ate anything at hand, either animal or vegetable. . . . Some flew in the air, some swam in the ocean, others strutted around on the land on their hind feet alone, while others walked on all fours. . . . They are not only the chief forms of life in the rocks where they are found, but some of

Fossils, Flood, and Fire

them far surpassed in size and strength any land animals now alive, and have been rivaled either in ancient or modern times, only by the whales.—G. M. Price, *The New Geology*, p. 519.

Some of the dinosaurs are worthy of special mention. *Brontosaurus* was about 65 feet long, and *Diplodocus* was nearly 85 feet long. Many of these creatures had a ridiculously small brain. *Stegosaurus* had in addition, in the sacral part of the spinal column, a large cavity, about 20 times the size of the real brain. This second "brain" evidently controlled by reflex action the movements of the rear portion of the body.

Ichthyosaurs (fish-reptiles) were nearly 30 feet long, and must have resembled a dolphin or porpoise. The skin was smooth, with no scales or horny plates. They had long, narrow snouts, and the jaws were armed with sharp teeth.

In contrast with these were the *Plesiosaurs*, with a long, slender neck and huge paddles nearly as long as the body, and a long tail that tapered to a point. One species was 50 feet long, over 20 feet of which was neck. The *Mososaurs*, closely related, grew as much as 100 feet long, and were snake-like in appearance—truly "sea-serpents."

Then there were the *Pterodactyls*, or flying reptiles. They had huge bat-like wings. The bones were hollow, like those of birds, and there was a true keel for the attachment of flying muscles. Some of them had a wing-spread of 20 feet.

It is a striking fact that, geologically speaking, all of the great reptiles suddenly became extinct the world over. That is, in spite of their abundance in the Mesozoic rocks, almost no sign of them is to be seen in the next series of rocks. This is one of the profound mysteries that popular uniformitarian geology cannot solve, and some writers are quite eloquent on this point.

A group of dinosaurs and related extinct reptiles. Drawings by Elden James.

Fossils, Flood, and Fire

While most of the invertebrates of the Colorado Plateau are non-marine, in some parts of the world extensive marine deposits occur in these rocks. Among the most abundant of all marine organisms in Mesozoic strata are the cephalopods. These are mollusks, some with coiled shells like the horns of wild sheep, some with straight shells, like unrolled coils. The curved ones are called *ammonites,* from the name of the ancient god Ammon, who was supposed to have had curled horns. The coiled ammonites vary from a few inches to large ones four feet across; the straight ones are six feet or more long.

Oysters are so abundant that whole reefs are built of their shells. Of the shrimps, crabs, and lobsters over 8,000 species have been described. Starfishes, sea urchins, and crinoids occur also.

The upper part of the Mesozoic, the Cretaceous, is much more widespread than either the Triassic or Jurassic. Geologists speak of it as a time of the "last great submergence." All the continents were awash, even to the interior in many places, in North America leaving only the Appalachian and portions of the Rocky Mountains above water. At this time great sheets of sediments from Alaska to Mexico were laid down, consisting of siltstones and sandstones, in contrast to the beds of dolomite and limestone of the lower rocks.

In the United States these sheets of sediments were of the nature of broad flood-plain deposits in the Rocky Mountain states, particularly on the east side of the mountains, which had just emerged from their original recumbent position. Detritus from the rising mountains spread far and wide, forming deltas and interbraided streams for hundreds of miles. In these deposits are found a marvelously rich and varied array of dinosaurs, more numerous and varied even than those of the famed Morrison formation. These dinosaurs were all plant-eaters. Associated with them were turtles, snakes, lizards,

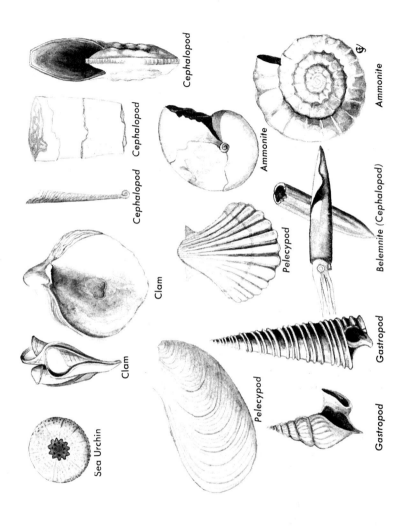

A group of Cretaceous fossils. Drawing by Elden James.

Fossils, Flood, and Fire

and crocodilians, frogs, toothed birds, and small mammals in great abundance. During the Mesozoic sedimentation striking changes took place. Whereas the eastern half of the continent had been covered with deposits coming from an unknown land on the east, now terrific over-washing was coming from the west. This resulted in the production of the "red beds" we have described in the previous chapters. In the Appalachian region great masses of material were removed from the uplifted sediments and carried eastward to form the present coastal plain reaching from New Jersey southward. Some sea life mixed with these sediments indicates that wave action was involved.

As we approach the close of the Cretaceous sedimentation worldwide tectonic movements are in evidence. The whole Pacific Coast from Alaska to the tip of South America rose as a great batholith, or plastic uplift, raising and folding all the sediments previously laid down along its course, and pushing up great masses of igneous rock to form the core of the Rockies and the Andes. The immensity of these movements is almost incomprehensible. The Front Range in Colorado and the Laramie Range in Wyoming have granite and other magmatic rocks rising to 11,000 feet, and some of the higher peaks of the Rockies rise 14,000 feet. The strata that had occupied the area now lie tilted against these granite cores and dip away in steep hogbacks. Thousands of feet of sediments have been washed off the top of these ranges and have been carried into the intermountain basins, and far out onto the plains.

Among other prominent features of the landscape produced at this time we might mention the Sierra Nevada batholith, the Idaho batholith, of which the Sawtooth Mountains are a remnant, the Boulder batholith in Montana, and the great Coast Range batholith in British Columbia. All of these were produced by molten or semi-molten material being forced up from

The Violent Climax

below. The Sierra Nevada uplift is about 400 miles long and 80 to 100 miles wide; the Idaho batholith covers some 16,000 square miles; the Coast batholith is 1100 miles long.

Uniformitarian geologists and diluvialists alike have been sorely perplexed over the question of how such tremendous movements of magmatic material could have taken place. Geologists have proposed many theories regarding plastic flow, convection currents in the semi-molten subcrust, crustal contraction, changes in speed of rotation, and others. Perhaps the most likely suggestion has come from the work of J. S. Lee, of the University of Peking, discussed in his *Geology of China*, 1939. He claimed that the outstanding tectonic features of the earth must have been produced by changes in the speed of rotation. No other known forces, he said, would be sufficient to throw the great highlands of the earth up for thousands of feet.

To demonstrate his theory, he coated drums with plastic substances on the inside to prevent their being thrown off by the rotational forces. By changing the speed of rotation, he produced patterns in plastic materials resembling the surface features of the earth.

The earth is 4000 miles from center to circumference. On the outside is a thin crust. By the time a depth of sixty miles or so is reached, the pressure due to the weight of the rocks is so great that all material becomes plastic. This is known as the zone of flowage. It would appear that the crust of the earth rests upon this plastic layer.

As long as the rotation of the earth is perfectly smooth, all will be well. But let the least amount of irregularity in the motion occur, either in speed or direction, and the rocks above the zone of flowage would be distorted. Movements of the magma would take place. This would account for the great batholithic uplifts. The location and extent of these would depend to a great degree on the nature of the original surface. It may be a significant point that all the great mountain regions of the earth occur along the lines of the ancient seas—geosynclines, the geologists call them. These were apparently the weak places in the crust.

With these principles in mind, we are now able to understand another class of crustal movements, the lateral displacements, or, as commonly called, overthrusts. For a classic example let us go to Glacier National Park in Montana, and see what has happened there.

The most noticeable feature is Chief Mountain, a block of limestone and argillite (slate-like rock) that stands out by itself on the east edge of the mountains. It is classified as Algonkian, which normally lies below the Cambrian, and does so in most parts of the surrounding country. But here it rests on Cretaceous rocks which extend eastward over the plains. In several places along the sides of the valleys south of Chief Mountain the same sequence may be observed as at this peak.

The feature that has given rise to much controversy is the low dip of the fault plane. It varies from level to 20 or 30 degrees. Some students of geology have found it difficult to believe that a mass of rocks could be thrust over others at such low angles. For this reason the whole overthrust theory has been challenged, and it has been asserted by some that the upper layers must have been laid down naturally.

On the other hand there are evidences that are impossible to apply to any other explanation than tectonic movements. In Marias Pass, where U. S. Highway 2 goes through the mountains, the contact between the Cretaceous and the overlying Algonkian Altyn limestone may be clearly seen. The Cretaceous is intimately contorted, crumpled, and broken. Some of the same phenomena may be seen just east of St. Mary Lake. The soft Cretaceous has been unable to stand up under

The Violent Climax

the pressure of the moving mass above it. The Altyn limestone is very hard, and has been little affected.

The effect of this great earth movement, commonly known as the Lewis overthrust, may be seen in various places in a region 15 to 30 miles wide and running from the middle of Montana as far north as Banff, Alberta. Throughout much of this region a belt of disturbed strata lies east of the main overthrust, in places extending as much as 20 miles out onto the plains. In many places the rocks are so much crushed and broken, sometimes overthrust, that the identity of the various formations is lost. The deformation decreases eastward.

South of Glacier National Park the disturbed belt is in the mountains, and is about 30 miles wide. The rocks are intensely distorted and broken. In many places are local thrusts.

Photographs of the higher mountains of the Park region show that contrary to the contention that the Algonkian has been laid down normally on top of the Cretaceous, in many places it has been twisted, folded, broken, set up on edge, or overturned. There is every evidence that astoundingly great forces have been at work. As one Park ranger once told me: "It must have been the most spectacular event the world has ever seen."

Space will not permit detailed discussion of the technical problems involved in these gigantic movements, but perhaps enough has been said to enable the reader to appreciate the terrific action involved in the formation of these mountains.

Somewhat similar effects may be seen in the vicinity of Banff, Alberta, where six mountain blocks have been tilted and shoved one over the other. Old mine workings on the east side of Cascade Mountain just north of Banff showed that the underlying beds were in-

Fossils, Flood, and Fire

tensely contorted, ground up, or completely metamorphosed where overridden by the thrust block.

To give detailed descriptions of the many examples of overthrusts given in geological reports would be far beyond the scope of this book. But anyone who has spent any time in the field will be convinced of the fact that stupendous distortional effects must have accompanied the closing paroxysms of the Flood.

Not only are these evidences among the strongest to be found for the Great Catastrophe, but they are of such enormous extent, and involve such incomprehensible forces that there seems to be no reasonable way to explain them otherwise. The idea that these mountains with their twisted, upturned, and overturned strata could have come about by slow, natural movements is impossible to believe.

One of the most outstanding features of the Cretaceous deposits is a wide belt of marls along the Atlantic coast. These are rich in potassium and iron, and are mined as fertilizer. In the West a similar feature is the extensive beds of potassium, classed as Jurassic, but practically of the same nature. These beds occupy great areas in northern Utah, and as far west as the Pacific coast. The origin of these potash deposits is somewhat of a mystery. Some geologists have suggested that they seem to have been derived from both the bones and the flesh of great multitudes of large animals. This explanation would seem to fit in well with the terrific disturbances that were going on, thus disintegrating and burying the bodies of large animals that had previously escaped the rising waters of the Flood.

At this time shifts in level appear to have produced a deep trough west of the Sierra Nevada batholith, and into this trough detrital materials from the rising mountains accumulated sandstone, shale, and siltstone to a depth of more than 50,000 feet. These sediments

The Violent Climax

occupied an area as far west as the present coast line in some places, and were later upheaved to form parts of the Coast Ranges of California, leaving a deep depression in the Great Valley.

While the Pennsylvanian rocks yield great quantities of coal and oil, the Mesozoic rocks are much richer in these products. Almost one-fifth of the world's oil fields are in Mesozoic rocks. The Middle East fields draw from Jurassic and Cretaceous. Coal is found in all the Mesozoic strata. Cretaceous ranks close to Pennsylvanian in coal production in the United States and Canada.

What has been said for the Mesozoic rocks of America is largely true for the rest of the northern hemisphere. In the southern hemisphere the deposits are so scattered and mingled with volcanic materials and erratic blocks that they are almost impossible to decipher. They do, nevertheless, contain an abundance of fossils similar to those of the north.

In China the great mountain-building movements seem to have occurred somewhat earlier in the Mesozoic series than in America. The Triassic is continental, and is often devoid of fossils. Intense folding and thrusting took place before the last sediments were laid down. The Jurassic consists of shales, sandstones, and conglomerates laid down in basins between the mountain areas that were being upheaved. It seems that in some cases we must imagine both uplift of the newly deposited sediments and some of the original land surfaces not yet destroyed by the action of the Flood. In much of China the Jurassic stage was one of violent upheaval and vulcanism. The Cretaceous of China was deposited only in continental basins and contains sandstones, mudstones, and papery shales. There is much lava, tuff, and many intrusions.

One of the best examples of the terrific tectonic ac-

Fossils, Flood, and Fire

tion that took place during Mesozoic "time" is seen in the great Rift Valley of Africa. Here is a great crack, as it were, in the earth's surface, running for a distance of 4000 miles, from Palestine down through the Red Sea and on into the lake region of central Africa. The bottom of the Red Sea, which is part of the trough, lies 7500 feet below sea level. Lake Tanganyika is nearly 4200 feet deep, and the cliffs bordering it rise 4000 feet above its surface.

The rift is a notable example of shattering on a world-wide scale, and it involved the crust of the earth in its whole thickness. It has been described as a virtual rending asunder of the crust. Geologists are at a complete loss to account for such a tremendous action to have taken place so near the close of the sedimentary series. Perhaps no other tectonic phenomenon gives such clear evidence for a sudden and catastrophic movement capable of tearing the crust apart along approximately one-sixth of its circumference. It was, as Suess once remarked, "the breaking up of the terrestrial globe." Geology underwent such changes that the surface of the earth would never again bear any resemblance to the original.

Not only do we have the great Rift Valley as evidence of tremendously powerful cataclysmic forces, but in a number of places smaller rift valleys were formed, resembling the African valley in all essential features except in magnitude. Recent studies have shown that there is a series of similar rifts along the swells, or "mountains" in the center of the Atlantic Ocean and in many places in the Indian and Pacific oceans. How these great earth movements can be interpreted on any kind of uniformitarian basis is hard to understand. Only in the light of a great world catastrophe can we gain any adequate conception of these terrific world-wide geological movements.

The Violent Climax

But the story is not yet complete. Striking as the evidence for the Flood shows up in the Mesozoic, we yet have one series of rocks for which we must account, the Cenozoic, or "recent-life" rocks.

CHAPTER THIRTEEN

Shaping the Landscape

By the time the Cretaceous uplift had taken place, the continents had assumed approximately their present shapes. It must not be assumed, though, that the action ceased suddenly. On the contrary it appears to have continued for a long time. It may be that crustal disturbances and tectonic movements took place for some time after the close of the Flood itself.

Let us now picture the surface of the earth as the great mountain regions have been forced up above the rest of the continental areas. In between these ragged masses would be deep basins, into which vast amounts of sand and silt would be deposited. It is largely in these basins that the next series of stratified rocks, the Tertiary, is found, although much of the same type of deposit was laid down out on the plains and along the ocean shores. The Tertiary and Quaternary, which were the third and fourth divisions of the original rock classification worked out between 1750 and 1830, make up what are called the Cenozoic.

These rocks were first studied in the Paris Basin, where French paleontologists noted that the beds contained many modern species of mollusks. Sir Charles Lyell proposed a scheme of classification based on the percentage of modern species in the various layers—a fact that seemed to be true in that region, although it has since been learned that it does not hold good elsewhere. But his names of the Cenozoic series are still used. They are as follows, reading from the top downward:

Pleistocene	(most-recent)
Pliocene	(more-recent)
Miocene	(less-recent)
Oligocene	(little-recent)
Eocene	(dawn-recent)
Paleocene	(ancient-recent)

The Tertiary classification is the most arbitrary in the whole scale. In fact it is a question whether various Tertiary deposits in different parts of the world can be correlated one with the other in matter of time, whether from the evolutionary viewpoint or from the viewpoint of the Flood. Classification is based largely on fossils, and the various Tertiary beds are scattered so widely, some in basins, some along the ocean shores, that those that contain similar fossils may not necessarily have been buried contemporaneously.

At this point the diluvialist is faced with a difficult problem: *When did the Flood come to a close?* Some Tertiary areas have marine fossils that indicate that the sea must have washed them into place, or they may have come from upland seas or lakes; some are remnants, or a continuation, of the Cretaceous types, so that the distinction between the Mesozoic and Cenozoic is hard to make. Others, such as in mountain basins, seem to have come after the uplifts had largely been accomplished. They contain the same types of life as exist today, together with many extinct types.

Shaping the Landscape

Some of the data assembled by the paleontologists may have value for the creationist who is trying to interpret the changes taking place after the Flood. For instance, many Paleocene animals appear to have been slow and clumsy. Perhaps this is why they were caught by the rising waters before some others were. Also, some mammals like the present, such as rodents, rabbits, etc., seem to have mingled with the others. But as we ascend into the Eocene, types more like our modern ones replace some of the others. It is an extremely difficult thing to trace the supposed evolutionary line from the lower levels to the upper. Why try? Why not recognize that there is simply the matter of different types, whether mingled in zones or assemblages or else being overwhelmed together, while some escaped or migrated in at a later date. In our present state of knowledge it is not easy to draw a line between the last Flood deposits and the post-Flood sediments.

One problem for the uniformitarian geologist is the fact that in the Tertiary new types of high taxonomic rank appear suddenly, without any fossils to show their ancestry. Could this fact not be interpreted to mean that they may have migrated into America after the Flood? Or could it mean that they lived independently in their own zones, without any kind of gradation with others?

A peculiar fact about the Tertiary deposits of the San Francisco Bay region is that in the mixtures formed by stratification, erosion, and eruption (including uplift and tilting of strata and the formation of some mountain ranges two to three thousand feet high) we find fossils of the same vegetation types living in the area today—types especially adapted to the peculiar climate of the region. It is hard to imagine the same climate there before the Flood as exists today, and so it seems as if these deposits must have been made after

Fossils, Flood, and Fire

the specially adapted types had had time to develop after the Flood.

The early Tertiary climates must have been much milder than the present, for forests of a tropical nature extended as far north as 50 degrees, that is, to the vicinity of Winnipeg. Studies in Europe, particularly, show a mingling of types, which became more and more distinctively separated as the climate became more severe. It was apparently a period of great stress, and without doubt changes in plants and animals, particularly plants, took place much more rapidly than they now do. Possibly the duration of violent earth movements may have been much different in different parts of the earth. Rocks classified as belonging to a certain "age" or "period" geologically, because of fossil content, may possibly not be correlated in actual time of burial. Each case must be considered on its own individual merits.

Now let us notice some of the Cenozoic beds and see what we can learn about what was going on while they were being buried.

The whole Gulf coastal plain is filled with Tertiary rocks, which are estimated to be perhaps seven to eight miles thick. Wells in Louisiana 25,000 feet deep have failed to penetrate beyond the Tertiary. A subsidence has caused these beds to dip toward the Gulf. They consist of poorly consolidated sands, clays, and silts, and are finer toward the south. Land and marine deposits alternate. Some geologists have described their deposition as a battle between torrential floods from the north and gigantic waves from the south.

The Appalachians, while they were largely above the sea since the time of the Pennsylvanian sedimentation, or else anything later than that has been removed, have been subjected to heavy washing. Their contours were smoothed and at the time the Cretaceous

Shaping the Landscape

upheavals took place in the West, they also were uplifted, folded, and eroded.

In the Rocky Mountain region geologists speak of a great peneplain. This was the old level of the sediments which was arched up as the great highland arose. In this process two significant actions took place:—(1) the highlands underwent terrific washing which resulted in their present rugged topography, and (2) the plains to the east were dissected by water rushing down from the mountains, and were then filled by the tremendous outwash from the high mountain masses. In Colorado marine fossils, "dated" as the most recent known on the continent, are interbedded with stream-laid Paleocene deposits. This suggests vigorous interaction between the sea and streams from the land.

Several geological features give us a picture of what was taking place at this time. The Green River Basin in southwest Wyoming filled with sediments until an opening appeared in the Uinta uplift to the south. Whether a natural channel through a low spot allowed the water to drain from the basin, or whether an earthquake crack started the present gorge, we will never know. Anyway, the gorges of the Green River witness to tremendous erosion.

The Green River Basin accumulated sediments, thinly laminated, to the depth of about 2000 feet. These may be clearly seen in many places, but one of the most famous spots is the deposit near the old railway station of Fossil, a few miles west of Kemmerer, Wyoming. Assuming that the layers represent annual deposits, it has been estimated that their deposition took 6,500,000 years. But here a perplexing problem arises. In these thinly laminated beds thousands of fossil fish have been found. The skeletons of the fish show that they must have been as thick as sediments that are supposed to have taken hundreds or even thousands of years to accumulate. How, in all reason, can it be as-

Confluence of the Green River and the Colorado River in Canyonlands National Park, Utah. Photo by Ernest S. Booth.

sumed that fish skeletons would lie there for that long before being buried? When dug out they are perfect, indicating that they were buried before any disintegration had taken place.

A somewhat similar problem exists in the Wasatch formation, which is said to represent 11,000,000 years of deposition. It has a great deal of coarse conglomerate. Imagine such violence continued for all that length of time!

The Green River shales offer another puzzle. They are carbonaceous, consisting largely of marl, with varying amounts of organic matter between the marl layers. Marl is a clay with lime derived from shells or other calcareous portions of animals. The fact that these beds contain thousands of thin layers of marl indicates that they must have been laid down by rapidly moving

Upper: Dead Horse Point near Moab, Utah, showing how the Colorado River has cut into the horizontal strata. Photo by Harold W. Clark. Lower: Goose Necks of the San Juan River in southern Utah, showing the horizontal strata exposed by the river. Photo by Ernest S. Booth.

Fossils, Flood, and Fire

waters, for no quiet settling would spread this sort of material over such a wide area so evenly. And even if it did settle in a basin, something of unusual nature had to occur to grind up the calcareous material and mix it with the clay before it was washed into the basin.

The "Laramide orogeny," or mountain building, that lifted the Rockies in the Cretaceous stage did not cease until about the end of the Paleocene sedimentation. Some areas arose while others subsided, forming basins, which then filled with the rapidly eroding materials from the higher portions.

In the South Park, in Colorado, sediments accumulated until the overflow was let out through the Royal Gorge, which eventually cut down to a depth of 1200 feet.

In some places, as near Florissant, Colorado, and in the John Day basin in central Oregon, the basins were surrounded by active volcanoes, from which finely divided dust settled and became an important factor in the perfect preservation of fish, insects, plants, and mammals.

In the Big Horn and Wind River basins in Wyoming thousands of feet of early Tertiary beds were laid down, and then material from them was washed away to form the later Tertiary deposits farther out on the plains. This movement must have been extremely rapid.

In this region one may encounter a phenomenon that has caused some perplexity to diluvialists. It is known as "superimposed drainage." It is apparent that when the rapid wash-off of the rising Rockies to the west spread eastward into the basins between the Wind River and Big Horn and other mountains, it actually buried ranges thousands of feet high and formed flat flood-plains above them, completely concealing them. Streams of great volume meandering over these plains

Tertiary deposits at Painted Hills, eastern Oregon. Photo by Harold W. Clark.

wore down to the tops of the mountains and continued to cut right into them. As they did this, they received side streams from the surrounding areas, eventually carrying away thousands of feet of the original floodplain and redepositing it farther to the east. When we look at these large streams cutting through the ranges, when a few miles away is what appears to be an easy way around, we are puzzled. But if we can reconstruct the sequence of events, the whole matter becomes clear. Obviously when the cutting began, the way around was not open.

Out on the plains the outwash laid down an intricate pattern of local formations of sandstone and shale. Streams from the mountains loaded with gravel, sand, and mud produced coarse, sandy deposits. The Tertiary beds of the High Plains are not continuous, but are a patchwork, local in character, due to the irregularly wandering streams. It is also a notable fact that the higher beds of the Tertiary are finer than the lower beds.

Typical Tertiary beds at John Day, Oregon (just west of Dayville). Upper photo by Harold W. Clark; lower photo by Ernest S. Booth.

Tertiary beds at John Day, Oregon (just west of Dayville). Upper photo shows mainly Miocene beds; lower photo shows mainly Oligocene beds. Photos by Ernest S. Booth.

Fossils, Flood, and Fire

Erosion occurring at a later time produced the wierd scenery of the badlands, such as is found in South Dakota.

In the Colorado Plateau the great thickness of strata from Devonian upward has undergone extensive stripping. The action evidently began at the time the mountains were being eroded to furnish materials for filling the basins, and continued for hundreds of years afterwards. As one goes southward across the plateau region, he can see the effect of this stripping, as layer after layer of sediment has been exposed, gradually leading downward until finally the Grand Canyon has cut down into the non-sedimentary Archaean rocks.

The tremendous volume of water necessary to accomplish this action was evidently furnished by violent and frequent, possibly almost continuous precipitation acting upon the soft sediments left by the Flood. There is little or no evidence that the waters of the sea were acting on these deposits during this stage.

At the time the Colorado Plateau and the Rocky Mountain region were being eroded, other areas nearby were being built up by volcanic action. The San Juan Mountains in southern Colorado came up as a great magmatic mass. Some of the peaks are over 14,000 feet high. Similarly the La Sal Mountains in eastern Utah and western Colorado were built up; they are still over 13,000 feet high. The whole Columbia Plateau, covering about 200,000 square miles in Idaho and adjoining states, became covered with thousands of feet of ash and lava, to an estimated 24,000 cubic miles. The most of this material came up through fissures and small cones, and without any question the Cascades arose at this time. This is evidenced by the fact that fossil plants in the lower beds of central Oregon are of the type that require a mild, humid climate, while those of the upper beds are similar to those that live there today.

Shaping the Landscape

There is no evidence of any high cones in the region except along the Cascades.

The picture along the whole Pacific Coast is so complex that it is extremely difficult to unravel. A succession of events occurred, alternating between upheaval of mountain ranges, deposits of sediments, and erosion. This action extended as far east as the Rockies, where the Absaroka Range in Wyoming was produced. Eruptive action laid down a large plateau just east of the present Yellowstone National Park, and built it up to a height of 12,000 or more feet. Then erosion and tectonic action tore it up and carried away much of the material. Today there stands a mountain range 170 miles long and 84 miles wide, a remnant only of the originally huge deposit. Vast eruptions took place in the Aleutians and long lines of volcanic mountains were formed from Alaska to Mexico. Along with the eruptives were extensive lava flows, rising ridges, and deeply subsiding basins. It was a time of violence beyond comparison.

The violence continued in places throughout the deposition of sediments classed as later Tertiary and Pleistocene. Similar action took place in Scandinavia and in the Alps. Most of the Himalayas were uplifted in the Tertiary and Pleistocene.

These facts raise again the question of where the Flood ceased. It would seem that we must recognize some, at least, of the early Tertiary as the results of the closing Flood action. The change in climate evident in the western United States, as we have noted in the case of the plants of Oregon, seems to have come as late as the Pliocene.

Putting all these ideas together, it would seem that the Tertiary must be recognized, not as a long time after the Flood in which the present rate or erosion and sedimentation was active, but a period of great violence. The early part may well be considered as the closing par-

Upper: Basaltic lava flow in Yellowstone National Park, showing columnar structure formed when lava crystalizes upon cooling. Photo by Ernest S. Booth.

Lower: Succession of lava flows in the Columbia Plateau of central Oregon. Photo by Harold W. Clark.

Volcanic breccia in the Absaroka Range of Wyoming, formed from eruptive debris. Photos by Harold W. Clark.

Fossils, Flood, and Fire

oxysms of the Flood, after which came a short period of continued violence. We have not attempted to unravel the complex history of this period. Doubtless not all areas "dated" as belonging to certain "epochs" can be considered contemporaneous, because the dating is made on the basis of the fossils. But with so many violent actions going on in different parts of the world, the correlation of fossil sequences with actual events in time may have little significance. It must be remembered that all "dating" of rock strata is based on the assumption of normal operations of natural forces; but when we view these problems in the light of the terrific action of the closing days of the Flood and the period immediately after, such dating has little meaning.

The fossils of the Tertiary are in many ways similar to those of the present. This is true of plants and animals as well. This may well raise a question. Does this not indicate evolution, with certain types dying out and others taking their place? Not necessarily. While, if evolution had taken place, this would be the natural thing to expect, it of itself does not prove that evolution did take place. If we consider that when the

Erosion in the Valley of 10,000 Smokes since 1912. Katmai National Monument, Alaska.

Eroded ash flow in the Valley of 10,000 Smokes, Katmai National Monument, Alaska.

Volcanic ash in the Valley of 10,000 Smokes, Katmai National Monument, Alaska. The section above the stratified layers is wind-deposited. The section below the stratified layers is the original flow. The stratified layers are from air fall due to explosions after the ash flow.

Fossils, Flood, and Fire

earth was created it was stocked with life consisting of assemblages of species, each in its own niche, both in the waters and on the land, then if we picture the rising Flood waters burying these assemblages one after another, the last to be buried would be the one from which modern stock would be expected to come, provided any of them survived the cataclysm. Thus the plants which were left near the top would be able to sprout and grow and replenish the devastated earth. Then, too, many of the sea creatures would likely escape death, and would be able to multiply. And so, we can say that the situation is exactly what we might expect after a great catastrophe had destroyed much of the life of the earth. In the case of the land animals, the problem lies between those that may represent the destruction of the last of the pre-Flood kinds and those that might be considered as migrants into the area after the Flood. It is not at all difficult to realize that animals that are found at lower levels were the more sluggish and the least likely to migrate upwards in an attempt to escape the rising waters. Most of the fishes buried in the lower strata were bottom feeders, or other sluggish types. On the other hand, in the Tertiary deposits are found many large whales. Because of their high degree of mobility, they could have survived longer than some other types. It is my opinion that the Tertiary may contain representatives of both pre-Flood and post-Flood types, but just where any particular species should be placed cannot yet be determined until we have more information on which to base a conclusion.

A good example of the problem may be seen in the North Horn formation in Utah, which consists of three parts. The lower contains bones of dinosaurs and others, including some mammals. The middle layer is barren. The upper section contains bones of mammals. The lower is "dated" as Cretaceous and the upper as Paleocene, simply because of the fossils; how-

Shaping the Landscape

ever, there is not the slightest physical evidence but that the whole formation was laid down in one continuous action.

The matter of migration from one continent to another, and the relation between the continents during the different "periods," geologically speaking, is fraught with much perplexity. However, when we attempt to build up a pattern to fit the Flood theory, certain facts can be recognized, and certain problems must be admitted, which call for further study.

South America and Australia seem clearly to have become separated from other land masses, and to have "developed" (if we may use that term without becoming involved in the evolutionary interpretation) distinctive types. In South America are the llamas and their type; in Australia are the kangaroos and their kind. How much variation was possible within each major type, no one can say with certainty, but it is evident that a considerable amount did take place.

Was the Bering bridge the only way animals could come to America? No one can say with certainty. Studies seem to indicate the possibility of a connection with Europe by way of Greenland and Iceland. Some feel that South America was connected to Australia by way of Antarctica. And we must remember that for a long time after the Flood the climate of the northern lands was not like that of the present. It has also been suggested that a land connection may have existed by way of the West Indies to southern Europe. Many changes have taken place on the surface of the earth since the Flood, and we cannot be too arbitrary as to what could or could not happen.

In passing reference should be made to the so-called "mammals series," that is, to series of elephants, camels, horses, or others, which one can see in natural history museums, all arranged in the supposed order of their evolution.

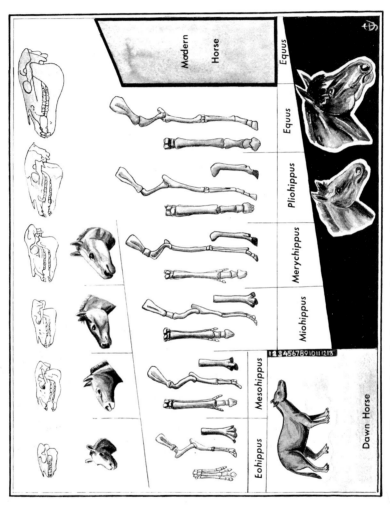

The evolutionists' scheme for the origin of the modern horse. Drawing by Elden James.

A skeleton of an extinct horse (*Equus shoshoniensis*) being excavated at Hagerman, Idaho. Leg bones can be seen in the center of the photo. Seven skeletons were found here, nearly all touching each other. Photo by Ernest S. Booth.

A partially exposed skull of the extinct horse (*Equus shoshoniensis*) at Hagerman, Idaho. Photo by Ernest S. Booth.

Teeth of three-toed horse, *Mesohippus bairdi*, found at John Day, Oregon. This photo shows the teeth just as they were discovered in the solid rock. Photo by Ernest S. Booth.

Two sets of three-toed horse teeth, *Mesohippus bairdi*, from John Day, Oregon, after they have been removed from the rock and mounted in blocks of plaster of paris. Photo by Ernest S. Booth.

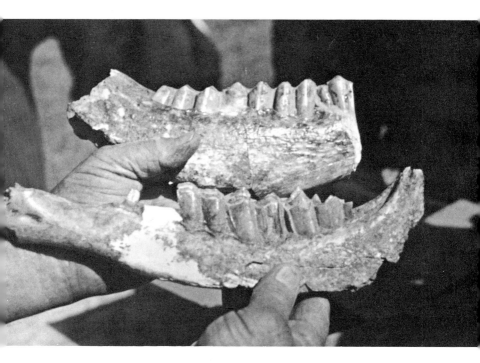

Jaws of extinct horses from southern Idaho. Photo by Ernest S. Booth.

At the right are the lower jaws of an extinct camel. At the left are two sets of teeth of the fossil horse, *Equus shoshoniensis*, all from Hagerman, Idaho. Photo by Ernest S. Booth.

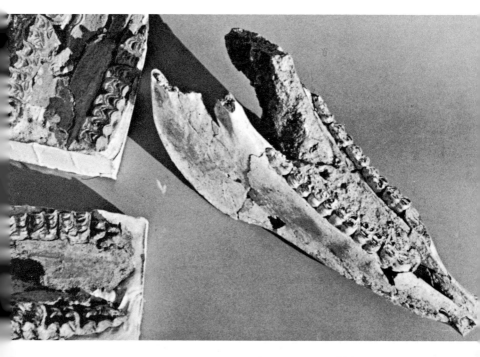

Fossils, Flood, and Fire

Perhaps the most striking of these are the horses and certain smaller types generally considered as their ancestors. These are found at different levels in the Tertiary deposits of the West, both on the plains and in Tertiary basins.

The smallest of the line is *Eohippus,* found in the Eocene. It varied in size from that of a cat to that of a small terrier. It had four toes, with the fifth one rudimentary on the front foot and the first and fifth rudimentary on the hind foot. The teeth had rounded cusps for grinding.

Orohippus had teeth like those of *Eohippus,* but lacked the rudimentary hind toes. *Mesohippus,* from the Oligocene beds, had three toes. *Parahippus,* in Miocene beds, had the outer digit much reduced, and the middle toe comparatively more prominent. The teeth showed the first indication of an open cement filling between the cusps, which is a feature in horses. *Pliohippus* was much like modern horses in both foot and tooth structure. A number of other forms might be listed, but these are sufficient to show the trends in the different beds.

These forms constitute what the evolutionists regard as a perfect evolutionary series, with a gradual succession of changes in size, foot bones, and teeth. The series is admirably arranged as we view it in a museum, and looks quite convincing. What has been said of the horse series might also be said for any of the others, such as elephants or camels. But is evolution the only answer to the problem?

The series arrangement may, partially at least, represent mere taxonomic series rather than descent lines. That is, any group of similar animals, such as horses, camels, elephants or others, may naturally be expected to show variation in any one or more structural features. The fact that two or more such structural features show somewhat parallel series of changes does

Shaping the Landscape

not prove them evolutionary. These features have an adaptive significance. An animal living in an environment that requires certain adaptive features in one organ or system will generally show corresponding adaptations in other systems of structural features.

To illustrate this point: A "horse" or horselike animal with teeth adapted to masticating grass would need the proper type of feet to carry him over grassy plains. An animal with teeth adapted for browsing might need different feet in order to get around in the forest.

Not all structural points can be shown to have adaptive significance. In some cases other principles may account for them. Ordinary mutations may account for some of them. When all that is known of an animal is a few skeletons, it is hard to tell whether the specimen represents an actual species or a mere variation within the species. One such mutant might add to the list of fossil "species." There is no proof for evolutionary sequence.

Many members of the horse series have been found in sufficient numbers to make it obvious that they were a *bona fide* series, but instead of constituting an evolutionary sequence they may have been ecological types. Three-toed horses, especially those with spreading toes, would be fitted to life in marshy or soft ground. Their teeth would have fitted them for browsing on twigs or coarse herbage.

The larger horses were fitted for life on the plains. The single toe, the long, slender leg-bones, the fusion of tibia and tarsus, all point to adaptation for speed on hard ground. The teeth are adapted for chewing the grasses that naturally grow on the plains.

Between the "marsh horses" and the "plains horses" are adaptations for conditions in between the two extremes. They are often spoken of as the "forest horses."

The three classes of "horses" listed, when viewed in the light of ecological principles, can be explained

Fossils, Flood, and Fire

without the necessity of resorting to evolution. And if they all existed contemporaneously, they could very well have occupied different zones in the years immediately following the Flood, or even before.

A candid review of the data cited will show that the whole idea of serial progression in horses, camels, elephants, or other types, is based on purely hypothetical concepts, the same as any other evolutionary theory. On the other hand, we may explain the series as taxonomic or ecological groups, or by considering the possibility of hybridization between closely related groups, without having to accept the commonly offered evolutionary interpretation.

CHAPTER FOURTEEN

The Reign of Winter

About 1820 European geologists became interested in the loose sands, gravels, boulders, and clay that formed hillocks rising above small lakes in scattered localities and in the alluvium along the banks of streams. In 1839 Lyell applied the name Pleistocene to this debris. A little later, when Louis Agassiz published his studies on the glaciers, these materials became accepted as the results of ancient ice masses. These apparently spread out from the Alps and Scandinavian mountains, and, joining with small ice caps over the British Isles, overran much of northern Europe. In America there were no high mountains in the northeastern portion of the country, from which ice could have come, yet the same phenomena were clearly to be seen north of a line running from Long Island to the Ohio River, thence down to its confluence with the Mississippi, thence up the Missouri River to the Rockies. Thus the American ice would constitute a real "continental" ice sheet. But whether in North America or Europe, and no matter in what regions these materials might be found, they appeared to have been formed after all other strata, and to overlie everything else except the alluvium of the river banks.

Fossils, Flood, and Fire

What were the evidences upon which the glacial theory was based? Let us review briefly the "glacial" phenomena.

The most noticeable phenomenon one sees in the mountains where glaciers have been at work, is the polish on the rocks. In Yosemite one may see acres of granite so smoothly polished that it reflects sunlight like a mirror. In other areas, where weathering has attacked the polished surface, it has been somewhat dulled, yet it remains a striking witness to the action of the ice that passed over it.

Closely associated with the polish are the striae, or scratches, on the polished surfaces. These are supposed to have been produced by sand grains pushed along or imbedded in the bottom of the glacier. Sometimes larger grooves occur. Their origin is not clear.

All mountain glaciers receive debris from the cliffs along their course, in the form of rock fragments of all sizes up to blocks as large as a house. These are carried by the slow-flowing ice stream and eventually dropped at its snout to form a moraine. If the ice is melting back faster than it is advancing, instead of a moraine, it will form a series of recessional ridges, or merely a lot of scattered boulders or erratics, resting on the polished floor where they were dropped by the melting ice. All of these features are uniquely glacial, and can be clearly distinguished from water-formed phenomena.

The rivers issuing from beneath the glacier snout carry away the debris dropped there, spreading it out in outwash. This may spread out in gravel plains or may form valley trains of boulders along the sides of the valleys below the ice.

All of these phenomena may be clearly observed over continental areas where ice is supposed to have been in the past, such as northeastern North America, northern Europe, and portions of Asia and South America. It is

Map showing maximum extent of glaciation in North America. Drawing by Elden James.

Fossils, Flood, and Fire

the almost identical evidence to those seen in mountain areas that has led geologists to believe that great ice masses existed on the continental areas; and there seems to be no good reason why diluvialists should not accept this evidence, providing, of course, that it does not involve time elements that are out of line with Bible chronology.

A number of other phenomena are found in the continental areas supposed to have been glaciated, and of these we must make mention.

Eskers are long, narrow winding ridges of rudely stratified materials. They are seldom seen in modern glaciers, and their formation has been the cause of much speculation. They appear to have been formed by deposits of sediment in tunnels in the ice. Related to them are *kames*, which are low conical hills of sand and gravel, believed to have been produced by streams pouring from the end of an ice mass or down into holes in the ice and dropping their loads.

One of the most puzzling phenomena is that of the *till*, or boulder clay. It is a stiff, tenaceous, unstratified clay, full of subangular stones of various sizes, and in all positions. These are usually derived from the underlying rocks, with some mixture of foreign materials. They are usually covered with scratches. The clay is hard and difficult to cut through. It is generally spread out thinly, often as a veneer that fills the lower depressions.

Referring back to the Tertiary rocks, it is a noticeable fact that the Eocene contains plants of a subtropical nature, world-wide in extent. The Miocene shows evidence of a cooling climate, but even at that redwoods and magnolias flourished as far north as Greenland, and a continuous forest encircled the entire Polar area. In the Pliocene plants we have evidence of a marked world-wide cooling. In America this is correlated with the "Cascadian uplift," that is, with the rise of the

Kettles and drumlins at Williams Lake, British Columbia. Photo by Ernest S. Booth.

Kettles are depressions in glacial outwash which often fill with water. Drumlins are smooth hills shaped like inverted bowls of teaspoons, and left by receding ice.

Kettles, drumlins and erratic boulders in the northern part of Yellowstone National Park, Wyoming. Photo by Ernest S. Booth.

Fossils, Flood, and Fire

Cascade mountains in the northwest. This uplift also affected the continent as far east as the Rockies, and elevated much of it to its present height.

We have already pointed out the tremendous amount of volcanic activity going on during the time of the deposition of the Tertiary rocks. It not only threw great quantities of dust and ashes into the atmosphere, but it effectually shut off the warm winds from the Pacific, with a resulting lowering of temperature over the interior of the continent.

The vulcanism theory of glaciation has been ably defended by competent authorities. The opacity of the atmosphere may be reduced by water vapor, carbon dioxide, and dust. There is every evidence that during the Tertiary "time" greater humidity prevailed over much of North America than at present. In times of high humidity half of the sun's rays will be absorbed by a cloudless sky, while a cloud layer will reflect more than 70 per cent of the sun's rays. Dust may reduce solar radiation received by the earth by as much as 20 per cent.

Not only are these facts well known to meteorologists, but it has also been shown that a shell of volcanic dust around the world would be 30 times more effective in shutting out the solar radiation than in keeping the heat of the earth in. This is known as the "inverse greenhouse effect." And so, when the volcanoes of the Cascades were erupting along with others strung along the length of the Americas, and were supplemented by those of Asia, it is obvious that an enormous amount of dust would have been thrown into the atmosphere. Of course if we were to think of this volcanic action spread out over millions of years, it would not mean much; however, if we consider it as all being concentrated into a few hundred or thousand years, then it would be tremendously significant.

Kames at Williams Lake, British Columbia. Photo by Ernest S. Booth.

Drumlins at Williams Lake, British Columbia. Photo by Ernest S. Booth.

Erratic boulders left by a glacier in Yosemite National Park. Upper photo by Harold W. Clark; lower photo by Ernest S. Booth.

Erratic boulders dropped by a glacier in Yosemite National Park, California. Photo by Ernest S. Booth.

Erratic boulders and glacial moraine in Yellowstone National Park, Wyoming. Photo by Ernest S. Booth.

Fossils, Flood, and Fire

Contrary to common opinion, a glacial period would not require long cold winters, but rather cool, damp summers. It has been calculated that a difference of 5° Fahrenheit from the present annual average, other factors being favorable, would be sufficient to bring on glaciation. Perhaps even more important than the temperature factor is that of moisture. In order for glaciation to occur, abundant precipitation would be necessary. This situation was apparently amply met in the years following the Flood.

Any great ice mass as it grows, will set up anti-cyclonic action at its center. An example of this is seen in the Greenland ice sheet today, which is overwhelmed by air circulation that spreads out from the center. Over the Antarctic ice mass there is the same circulation, but of greater intensity. In the Greenland glacier, the wind blowing outward for most of the year, sweeps snow off the margin for about 50 miles. Back of this margin the ice is stagnant. Perhaps these facts may aid us in forming a picture of what may have been the condition when the great ice mass extended over much more of the continent than at present.

As the ice mass accumulated, it would chill the nearby air, causing an increased precipitation along the northern edge of the storm belt that would lie south of the ice. The result would be the accumulation of immense amounts of snow and ice over the northern portions of northern America and Europe, without greatly reducing the temperatures immediately south of the ice.

It has generally been supposed by diluvialists that the ice of the northern regions came on suddenly as the result of atmospheric changes at the time of the Flood. The frozen mammoths of Siberia have been cited as evidence of this. However, recent studies have shown that this is not a valid conclusion. Of the mammoths we shall speak a little later. But as to the suddenness of the change, it must be remembered that

Glacial moraines and erratic boulders, Yellowstone National Park, Wyoming. Photos by Ernest S. Booth.

Moraines are deposits of rocks, gravel and finer sediment left by a glacier. A lateral moraine occurs alongside the glacier whereas terminal moraines occur at the snouts of glaciers.

Lateral moraine at Athabaska Glacier, Jasper National Park, Alberta.

Williamson Rocks near Anacortes, Washington, showing planing action, polish, and grooves carved by a glacier formerly covering most of western Washington in the lowlands. Photo by Ernest S. Booth.

when the earth emerged from the Flood waters, it would still be at a mild temperature. The vast amounts of water in the oceans, having recently come from the interior of the earth, would be warm. (Diluvialists do not ascribe the bulk of the present waters on the earth to the 40 days of rain, but to waters coming from the earth itself, as the Genesis record suggests in the expression "the fountains of the great deep.") It would take a long time for the oceans to cool off. Competent scientists have estimated that it would require at least two turn-overs of the ocean waters, perhaps taking about 1000 years. Evidences from Siberia indicate that the northern limit of trees was once several hundred miles farther north than now, indicating a former milder climate.

Some diluvialists have questioned the possibility of thick accumulations of ice, as it has been said that the ice would melt of its own weight. But recent studies have thrown new light on the problem. Instrumental

Mountain planed and polished by glacier at Tuolumne Meadows, Yosemite National Park, California.

Mountain rounded, planed and polished by glacier at Tuolumne Meadows, Yosemite National Park, California. Photos by Ernest S. Booth.

Fossils, Flood, and Fire

measurements have shown that the Greenland ice cap averages 5200 feet in depth, and in places is 10,000 feet thick. Other ice masses are known to be thousands of feet deep. Laboratory experiments on ice have shown that if it is cooled to 27.5° Fahrenheit, it could accumulate to a depth of over two miles before it would melt of its own weight. If cooled to 21° it would require four miles in order for its pressure to melt it. Now it is recognized, of course, that when ice freezes, it is at 32°. But if it stands for hundreds of years, exposed to the cold of Arctic or Antarctic regions, it certainly will be chilled below the temperatures mentioned, and therefore would be capable of piling up to great depths. Even far away from the centers of greatest cold, a large amount of cooling would undoubtedly take place.

Modern glaciologists do not believe that ice can do much erosion. Its chief action seems to be in polishing and scratching its floor by means of the sand carried beneath it. However, it must be recognized that ice under great pressure thousands of feet thick must exert tremendous force on its bed, and could pry off loose fragments of rock against which it might come in contact.

Geologists have discovered such variations in the glacial deposits that they have postulated four different glacial periods in the Pleistocene, with warm periods in between. In America these are named from the bottom up, the Nebraskan, Kansan, Illinoian, and Wisconsin stages. Similar stages are named in Europe. This hypothesis is based on the finding of beds of peat and forest materials beneath the glacial gravel. The first three stages are called the older drift, in contrast with the Wisconsin, or newer drift.

The older drift lacks topographic expression, but is spread out thinly and widely, and has been so eroded and covered with wind-blown loess that its indentification is exceedingly difficult. It lacks morainic charac-

Glacial grooves, striae and polish just below Athabaska Glacier, Jasper National Park, Alberta. These were formed during the past four or five years.

Geology students examine glacial grooves, striae and polish along the highway near Banff, Alberta. These were formed perhaps 200 to 300 years ago. Photos by Ernest S. Booth.

Glacial polish on granite rock. Yosemite National Park, California. Photos by Ernest S. Booth.

Glacial grooves, striae and polish in granite rocks. Ontario, Canada. Photos by Ernest S. Booth.

Fossils, Flood, and Fire

teristics. Pebbles contained in it are often well rounded, suggesting water action. As a rule it is weathered more than the newer drift, and this has been taken as evidence for a longer time of exposure to the elements. However, when we realize that after the Flood the climate was semi-tropic quite far north, we can understand how some materials left over from that catastrophe might show a high degree of weathering. It may be possible that some of this material represents debris of that nature, although probably not entirely. There are a great many assumptions on both sides of the question, and it is not wise to be too arbitrary in regard to possible interpretations.

It has been seriously proposed by students of glaciation that the whole idea of multiple glaciations be abandoned in favor of the idea of one glaciation. The materials that have been explained on the basis of the supposed successive glacial stages may be only the outwash from the Wisconsin stage, or possibly some may be due to floating ice from the ice front. On this point we read from the pen of an advocate of diluvialism:

> If this concept is accepted, and it certainly seems to be supported by much evidence, there must be a revolution in geological thought. . . . It obviously correlates with the concept of post-Deluge affects which we have been advocating.—Morris, Henry M. *The Genesis Flood*, p. 302.

In contrast with the older drift, the newer, or Wisconsin, shows strong glacial topography—eskers, kames moraines, outwash plains, striations, etc. Water-sorted material is abundant, and little denudation has taken place since its deposition.

The present course of the Missouri River is an example of the way in which ice masses changed river courses. It is noticeable that as we approach the Missouri from the west, we pass over plains that are free from glacial debris, but as soon as we cross the river,

The Reign of Winter

glacial topography, boulders, and piles of debris are everywhere. As the Missouri comes eastward from Montana, it should, by all appearances, flow down the slope to join the Mississippi River in Minnesota. Instead, it turns southward abruptly, and goes through extensive hills. There is nothing in the topography to cause this change. We can think of no way that it could have come about except that its flow was diverted by the ice mass.

The Ohio River shows similar conditions. South of it are few if any glacial deposits, but north of it is a succession of moraines and other ice-formed materials, as well as shore lines of ancient lakes that occupied temporary basins until the ice mass melted away and allowed the present Great Lakes to form.

As one flies over the area he can plainly see the many old stream canyons coming from Kentucky northward toward the Great Lakes. But the Ohio River has captured them. Like the Missouri, there is nothing in the topography to explain how the Ohio took the course it now follows. Ice action seems to be the only possible explanation.

Another effect of the Wisconsin ice was the formation of ancient Lake Agassiz. In northwestern Minnesota, northeastern North Dakota, and in Manitoba the prairie floor is so flat that it gives every indication of having once been the floor of a lake. But there is no high land to the north to create a barrier to form such a lake. Furthermore, there is clear evidence that the Red River once flowed southward across the divide between its present course and the Mississippi drainage. The most reasonable explanation is that the ice blocked the northern end of the Red River valley and formed a great lake—Lake Agassiz. When the ice melted away, the drainage was established to the northeast to Hudson's Bay.

In somewhat similar manner the ice created marked

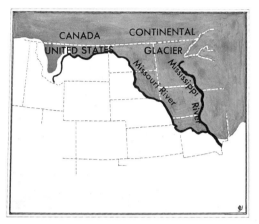

Map to show how the continental glacier changed the course of the Missouri River, diverting it from its natural drainage into Hudson's Bay, Canada, southward through an area much higher in altitude, where it flowed along the edge of the ice until it reached the Mississippi. The change of direction took place in northern North Dakota. Drawing by Elden James.

changes in the drainage of the Great Lakes System. There is no reason why the diluvialist may not accept the descriptions in geology books as valid, even though we cannot go along with the long periods of time commonly believed to have been involved.

The high mountains of the West show glacial debris that indicates action in the past of much greater magnitude than anything now going on. In Yosemite the ice came down to the lower end of the valley, and from here up into the mountains and to the tops of the peaks such as Lyell, tracings may be seen clearly. The mountain glacial phenomena, however, are open to the same questions as are the glacial evidences on the plains.

One of the most perplexing phenomena of geology is that of the so-called "frozen mammoths" of Siberia. Since the 4th century large numbers of tusks have made their way into the Chinese and European markets, and fantastic tales have grown up as to their source. Some of these tales have been repeated in all sincerity by

Banks Lake in Grand Coulee, eastern Washington. This is the former channel of the Columbia River during the "ice age" when a glacier dammed the river at the present site of Grand Coulee Dam. The river flowed for some years through this channel until the ice melted, then the river went back to its original channel.

Dry Falls, eastern Washington, formerly the world's largest waterfall. During the time when the continental ice mass covered much of northern North America, ice dammed the Columbia River near Coulee Dam, sending it south through this channel for a number of years. Photos by Ernest S. Booth.

Mountain glaciers always carve U-shaped valleys. The upper photo is Yosemite Valley in California while the lower photo is a glacial valley near Smithers, British Columbia. Photos by Ernest S. Booth.

A typical cirque, or rounded pocket formed in the side of a mountain by a former glacier. As the ice pulls away from the mountain it removes boulders constantly until this depression has been dug into the side of the mountain. Such cirques can be found in mountains as far south as New Mexico and southern California. This one is near Prince Rupert, British Columbia.

A typical glacial lake formed after a glacier has melted away leaving this depression in the mountainside which then fills with water. This lake is near Smithers, British Columbia. Photos by Ernest S. Booth.

Mt. Athabaska, Alberta, Jasper National Park, with one of its many glaciers.

Athabaska Glacier, Jasper National Park, Alberta, showing the snout of the glacier at the right and the lateral moraine at the left. Terminal moraine material can be seen in the foreground. Photos by Ernest S. Booth.

The surface of Athabaska Glacier, Jasper National Park, Alberta, showing "ice falls" in the background.

The snout of Athabaska Glacier, showing morainal material at the terminus. Photos by Ernest S. Booth.

Peyto Lake, Banff National Park, Alberta, showing glacial outwash at the left of the lake, and a glacial stream flowing into the lake from Peyto Glacier to the extreme left and out of the picture.

Glacial lake at the terminus of Athabaska Glacier, Jasper National Park, Alberta. Note the moraine in the background just above the lake. The darker color in the lower portion of the moraine is due to melting ice underneath the rocks of the moraine, for this moraine covers a large part of the glacier. Photos by Ernest S. Booth.

Two striking aerial views of Kishkawalsh Glacier, in the St. Elias Range, Yukon Territory, Canada. Note how morainal material falls onto the glacier from the mountains and is carried along by the ice for a number of miles down the valleys. Note in the lower photos that two glaciers are fusing together in the lower part of the valley, yet there is no mixing of ice or morainal materials from the two.

Icebergs from Portage Glacier, Alaska, as the ice from the glacier breaks off into the ocean.

Two glaciers in the St. Elias Range, Yukon Territory, showing how glaciers produce moraines along the edges of their flow.

Placer mining with hydraulic pressure uncovers gold near Fairbanks, Alaska. Gold is found in these extensive beds of frozen muck. This muck also contains fragments of flesh and bones of many prehistoric animals which apparently were killed at the onset of the glacial period.

This view shows details of frozen muck and ice lenses in the same area as the photo above.

Fossils, Flood, and Fire

scientific writers, who have not critically examined their validity. Since some animals have been found in the mud of riverbanks, the natives of Siberia called them *mammut,* or ground-hogs, believing that they burrowed in the soil. The word does not indicate that they were mammoth as far as size goes. In fact, the Siberian mammoth was about the size of the modern Indian elephant.

Both diluvialists and evolutionists have been much confused as to how the facts regarding the mammoths should be interpreted. In 1887 Sir Henry H. Howorth wrote a large volume entitled *The Mammoth and the Flood.* In it he attempted to show that thousands or millions of these animals were overwhelmed by a sudden catastrophe. He told of great quantities of bones along the Arctic shores of Siberia, in fact, whole islands of almost nothing but mammoth bones. Bones piled up at the mouths of some of the rivers like driftwood, he declared. He said that the mammoths were frozen so suddenly that their flesh was preserved so well it was palatable to both dogs and man.

Some creationist writers have followed Howorth's ideas without critically checking their sources. But in recent years it has been realized that his conclusions were not supported by sound evidence. He seemed not to have been careful to distinguish between facts and wierd tales told by men who either stretched the truth for the sake of making tales, or who spoke from imagination, without much background to make their stories worth while. In fact, a few years ago when I tried to run down some of the material in Howorth's books, I found it extremely difficult to do. They seemed to be composed of a mass of clippings thrown together by incompetent secretaries, with no real factual matter to back them up.

In our day one very popular writer has resorted to startling and spectacular portrayals of the overwhelming catastrophe which brought air down from the upper

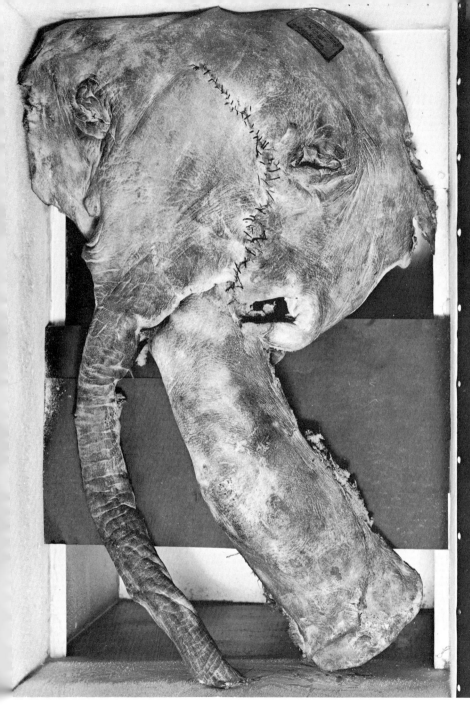

This young woolly mammoth was found in Alaska in frozen ground. It is intact, with skin, hair and flesh over the bones, just as it was frozen centuries ago in the arctic. It is on display in a deep freeze at the American Museum in New York. Photo courtesy of the American Museum.

Mounted skeleton of a mammoth in the Los Angeles County Museum. Mammoths such as this roamed North America before the coming of the glacial period, most likely during the glacial period, and possibly for a long time after the glacial period, for remains have been found in many places just below the surface of the ground. Photo by Ernest S. Booth.

atmosphere at a temperature of 150 degrees below zero, in order to freeze the mammoths so quickly that disintegration would not take place.

Now what are the real facts about the mammoths?

In the first place, the number of frozen giants has been grossly over-estimated. Instead of millions or even thousands, the fact is that in 250 years only 34 such specimens have been found, and many of them consist only of fragments. These have been hard to find; many reports have been checked by the Russian government, but without success.

As to the flesh being fit for food, good authorities declare that the idea is pure fiction. The truth is that the flesh is putrid, with a horrible odor, and the tissues are so badly disintegrated that their microscopic structure cannot be observed. In fact, even gross anatomical studies are almost impossible, the flesh is in such a poor state of preservation. Claims that the flesh has been fed to dogs may have some validity, however.

The Reign of Winter

But now what about the general belief among diluvialists that these animals were overwhelmed by the Flood?

The Arctic islands of both Siberia and Alaska contain fossils of warm-climate plants and animals, such as might have existed before the Flood. Marine specimens are common. On the other hand, plants and animals in deposits where mammoth bones are found are similar to those existing in the region today. Furthermore, the mammoths are known to have had heavy fur, which would have enabled them to live in a cold climate. They had a heavy coat of subcutaneous fat, also.

Most of the bones of mammoths are associated with bones of many other animals such as live in northern regions today. They are found in alluvium, or river-deposited silt, which varies from 30 to 100 feet thick. This alluvium is said to contain several layers, indicating a succession of deposit. It seems to have been produced by the erosion of the surrounding hills and mountains, producing material which was carried out over the vast plains that occur in Siberia and Alaska. And—note this—in this material there is not anything like the density of bony material that popular writers have pictured.

When we gather all these facts together, it appears quite evident that the mammoths lived naturally in the northern lands for hundreds of years *after* the Flood. The fact that remnants of tree growth have been found 500 miles north of the area where most of the mammoths have been found in Siberia, would indicate that a change of climate some time after the Flood not only caused the tree line to be moved southward, but was very possibly concerned in the extermination of the mammoths.

One of the arguments put forward in regard to the length of time necessary for the formation of the glacial deposits is that from varves. If we go today to the head

Fossils, Flood, and Fire

of Lake Louise, in the Canadian Rockies, and note where the stream enters the lake, laden with glacial "flour," fine sediment scoured from the rocks by the glaciers in the valleys above, we can see that layer after layer of fine mud has been laid down as these sediments drop to the bottom. The material deposited in summer is coarser and thicker than that deposited in winter. Thus the varves, or layers, appear to be annual layers. Now, on this same line of reasoning, clays found in ancient glacial ponds or lakes seem to be varved, or layered, and from these varves the supposed age of the deposit is estimated. By this method, for instance, it has been estimated that it took 10,000 years for the ice to melt back from Hartford, Connecticut to Montreal, Canada. The varves in the different beds are compared, and a system of correlation established, by which it is claimed that they represent a sequence, one after the other. The same is done for European glacial varves.

But, this method is not accepted by all geologists. Some good authorities on glaciation claim that the melting of the ice took place simultaneously all over New England. The methods by which a sequential arrangement of the varves has been made, it is pointed out, are too uncertain to be valid. Studies on European lakes show that local conditions may cause a layering much faster than annually. Sometimes layers are produced in one night by a severe storm.

Much more detail might be given regarding all phases of the glacial problem, but when we put all the facts together, it appears that all of the true glacial action could be compressed into a comparatively short period of time. There is an increasing wealth of information that would lead us to believe that the earth warmed up suddenly after the glaciation. And so, the commonly accepted picture of thousands of years for an ice age must be revised. Without doubt some parts of the country were in the grip of the ice a compara-

The Reign of Winter

tively short time ago, possibly as late as the time of Christ. Probably the Scandinavian ice sheet lingered over the mountains until into the early historical times. In fact, Norse legends state that the land was almost uninhabitable when first entered.

CHAPTER FIFTEEN

Cave Men and Stone Ages

We now come to one of the most intriguing of all geological puzzles, that of prehistoric man.

About 1750 the suggestion was made that man had come up through a series of steps in civilization, passing through a stone age, a bronze age, and an iron age. Then, when explorations were made in caves and on river terraces in the 19th century, interest in this theory was revived. In 1824 Buckland published a work on the relics of the Flood, treating of mammals bones in particular. The next year a cave was explored which contained remains of extinct animals. In 1828 human bones were found in France, associated with flint implements and the remains of the mammoth, woolly rhinoceros, cave-bear, cave-hyena, and others. Other caves were found in various parts of Europe, with much the same remains buried in the debris on their floors.

In 1869 the Paleolithic, or Old Stone Age was set up, consisting of five divisions, each determined by the type of stone implements found therein. In dating these divisions, much dependence was placed on the fact that the rivers of Europe have cut from three to six terraces in the limestone, indicating different stations of river

Fossils, Flood, and Fire

level, the high terraces, of course, being the oldest. Human remains and artifacts have been found, not in the rock from which the terraces were cut, but in the deposits of sand and gravel on the terraces.

Another important feature is the presence of glacial moraines. These show that at one time the glaciers of the Alps extended as much as forty or more miles from their present terminations. It was at this time, doubtless, that they made the gravel deposits on the high terraces. The river levels stood from 100 to 200 feet above the present level. As the glaciers receded up the valleys, the rivers continued to cut into the moraines left by the ice. Both actions were involved in the deposits of gravel on the terraces.

Few, if any, evidences of human life are found on the high terraces, except for an occasional flint tool. Some of the human artifacts are found on the gravels deposited on top of the high terraces, but the greater number of them are found on the lower terraces, indicating quite clearly that there was not much human occupancy of the country until after the great bulk of the cutting had been completed.

In addition to the terrace deposits, human remains are found in caves that were cut into the limestone rock along the sides of the river valleys. Doubtless these caves were cut out by high water, and in later years as the river levels were lowered, man took advantage of the caves and used them as shelters. Apparently the depths of the caves were not used, but only the entrances. Fires were built near the openings, not very far in, in order to allow the smoke to escape. Probably the caverns were used to protect man from rain and wild beasts, and for storage of baggage and implements of the chase. Probably the deeper caverns were too damp for use.

Reference has been made to the fact that the onset of glaciation must have been some hundreds of years after the Flood. The evidence of this in Europe is clear

River Vezere, France, with caves of prehistoric man. Such caves are clearly post-Flood, since pre-Flood rivers could not exist through the Flood. Photo Courtesy of the American Museum of Natural History, New York.

when we study the fossil faunas there. Several distinct types have been recognized that are quite different from what may be found there today. They are:

1. The *warm plains fauna* of northern Africa and southern Asia. This includes 14 or more species, such as the elephant, rhinoceros, and hippopotamus.

2. The *temperate meadows and forest fauna* of Europe and Asia. Twenty-six or more species include such animals as deer, moose, bison, wolf, and bear.

3. The *high, cool mountain fauna,* with chamois, ibex, and ptarmigan.

4. The *steppe and desert fauna,* with species of horse, antelope, and jerboa.

5. The *tundra fauna,* including reindeer, musk-ox, and arctic fox.

Fossils, Flood, and Fire

From these facts it seems quite obvious that the climatic zones were once much more compressed than at present. That is, there must have been cold areas in the mountains at the same time that the climate of the valleys was warm enough to support animals that are now found only in warm climates. As the climate of Europe became colder, many of these types became extinct.

The same situation prevailed with respect to plant life. We find mixed together such semi-tropical types as tulip-trees, sassafrass, and locust, along with more temperate species such as oak, beech, willow, and larch. Bamboo, palm, bay, and gum trees grew along the shores of the Mediterranean.

The supposedly oldest human remains (until recent finds in Africa) were those discovered in Java in 1893 and 1894. While excavating on a river terrace workmen discovered an upper molar tooth, and about three feet away, the upper part of a skull. About 50 feet from this location they found a left thigh bone, and assumed that it belonged to the same creature as the tooth and skull.

Measurements showed the skull to be smaller than an ordinary human skull, but larger than that of an ape. The new animal reconstructed from these fragments was named *Pithecanthropus erectus,* the erect ape-man, and it was dated at about 500,000 years ago. The skull receded sharply from the brows, and over the eyes were prominent brow-ridges. These features led to the supposition that the creature was in the line of descent from apes to man.

Hundreds of articles, books, and papers have been written, and writers have violently disagreed as to the interpretation of these remains. At present, however, a number of similar remains have been found, and it appears that they are representative of a race that once ranged over much of southeast Asia.

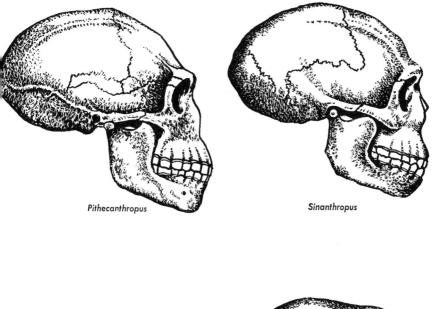

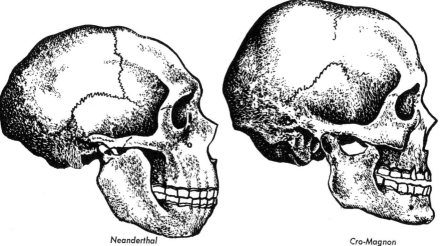

Skulls of common forms of fossil men. Java Man (*Pithecanthropus*) and Peking Man (*Sinanthropus*) are now recognized to be similar to modern man (*Homo sapiens*), and therefore are placed in the same genus (*Homo*) with modern man. Because Java Man and Peking Man resemble one another so closely they are placed in one species, *Homo erectus*. (See William W. Howells in "Homo Erectus," *Scientific American*, November 1966, Vol. 215, No. 5, pages 46-53.)

Fossils, Flood, and Fire

A massive jaw-bone dug out of a sand bank near Heidelberg, Germany, and named *Homo heidelbergensis,* is so fragmentary as to be of no practical value. However, it is dated at about 200,000 years ago. Still another worthless find was that of the Piltdown skull, found in England, which has proved to be a hoax.

The best known of all human remains are those of Neanderthal man, first discovered in the Neander Valley in Germany in 1857. Since then about a hundred specimens, mostly fragmentary, have been found, although a few of them are sufficiently complete to give a good idea of the anatomy of this race. They are distributed all the way from Belgium to Iraq and down into northern Africa.

Neanderthal man was short and robust, with a small cranium, but with limb bones much like those of modern man. He had prominent brow-ridges, and a sloping forehead. The arms were short, and the fingers short and stubby. While in many ways closely resembling modern man, yet he is said to have some features that gave a resemblance to an ape. In spite of the many illustrations showing him as stooped, it is now known that his posture was erect.

Early anthropologists thought that Neanderthal man was an ancestor of modern man, but they now believe that the so-called "primitive" features are secondary, that is, are due to an aberration, or departure from normal.

It has been said that if a Neanderthal man were to be dressed in modern clothes, given a hair-cut and shave, he would pass down one of our city streets with barely a notice. It seems to be an obsession with anthropologists to make their restorations of prehistoric man look wild and beastly. Much of the resemblance between their restorations and the apes is due to their vivid imaginations. When we get down to the solid facts, stripped of all bias toward evolution, the resem-

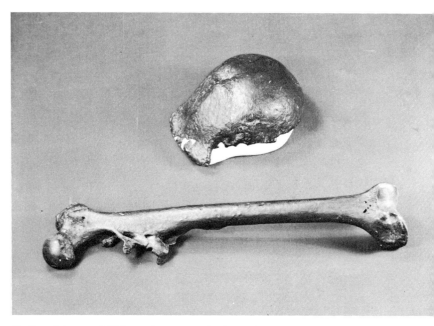

Skull and femur of Pithecanthropus, or the Java Man. Courtesy of the American Museum of Natural History, New York.

Skulls and reconstructions of Neanderthal Man. Courtesy of the American Museum of Natural History, New York.

Fossils, Flood, and Fire

blances are not nearly so great as has generally been supposed.

It is not clear whether they were killed off by invaders, or whether they had disappeared before Cro-Magnon man came into Europe. It is now generally supposed that they are merely the western periphery of a variation within the human species. Even the distinctive name, *Homo neanderthalensis,* may be misleading.

These early inhabitants of Europe were workers in flint that came into the region before the glacial period. There is abundant evidence that they arrived before the lowest descent of the ice, for caves have been found in which their remains have been buried by the advance of the ice.

In many places in central Europe evidence has been found for another race that appears to have followed the Neanderthals. This is known as the Cro-Magnon race. They were well built, identical to modern man in appearance.

Deposits where Cro-Magnon remains are found contain bones of horses, reindeer, wild cattle, and mammoth, together with all kinds of implements. In one camp on a river terrace the bones of horses were so numerous that it is estimated that 100,000 specimens must have been killed there for food.

One of the most striking records of the Cro-Magnons is found in the art work on walls of caves and on fragments of bone and horn. On the walls of limestone caves are found drawings and engravings of the woolly mammoth, woolly rhinoceros, cave-bear, bison, reindeer, lion, horse, fish, cattle, and stag. Carvings of these animals and of human figures were made on horn, tusks, and bone. These people evidently were a highly artistic race. Some of their drawings are superior to many modern pieces of art.

The transition from these people to those of historical times is well illustrated in the cave deposits. In a

Restoration of a Cro-Magnon scene with men painting pictures of woolly mammoths on the walls of the Cave of Font de Gaume in France. Photo courtesy of the American Museum of Natural History.

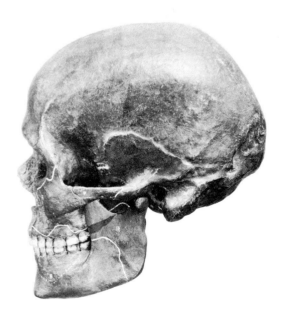

Skull of Cro-Magnon man. Photo courtesy of the American Museum.

Fossils, Flood, and Fire

cave at Mas d'Azil, on the French side of the Pyrenees, deposits 26 feet deep showed nine distinct layers. In the bottom layers were found gravel with fire-hearths, and harpoons of horn, flint, and bone. Higher up were found pottery, stone implements, and still higher, copper and bronze vessels. Finally near the top were Roman remains of iron, glass, and pottery. These layers reveal clearly that flint workers were followed by a race who used bone implements, and they in turn by the Romans.

Animal remains in these caves show that many of the animals of the lower layers disappeared and were followed by more modern types, similar to those found in the region today. Some of the upper layers contain bones of modern domesticated animals.

In recent years interest in prehistory has shifted to Africa, where a number of interesting discoveries have been made. In 1924 Raymond Dart found portions of a skull at Taungs, Bechuanaland. It was named *Australopithecus*, "southern ape." From 1937 onward many remains of this type were found in caves in Transvaal. Some slightly larger forms were named *Paranthropus*—"next to man."

In recent years critical analysis of these finds has placed them definitely as human, not apes. The skulls have several features not common to apes. The pelvic bones are constructed on the human plan. The canines are small, not overlapping as in apes. Thus, in spite of the fact that early reports assumed that these specimens were related to apes, and constituted a supposed evolutionary sequence, more recent conclusions by competent authorities are that they are essentially nothing but variations of the degree of racial difference, and not of separate species. At present the world is excited by reports of discoveries made by L. S. B. Leakey in Olduvai Gorge in Tanzania (formerly Tanganyika). In 1959 he unearthed an almost complete skull, which he named

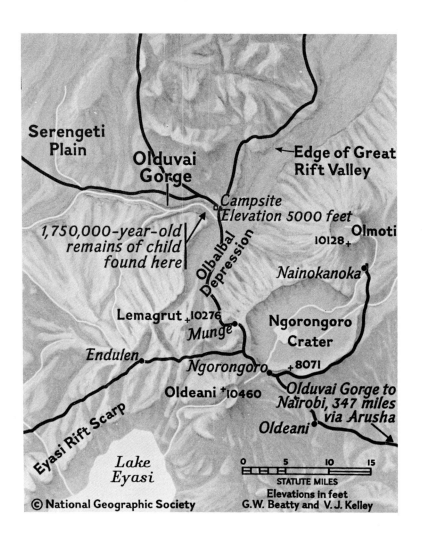

Map showing the location of Olduvai Gorge, where Dr. Leakey uncovered prehistoric men. Courtesy, National Geographic Society.

Dr. Leakey supervises the excavation of a giant elephantlike animal in Olduvai Gorge. Photo by Des Bartlett and Armand Dennis. Courtesy, the National Geographic Society.

This is Olduvai Gorge in Tanzania, east Africa. Dr. Leakey and associates stand on bluff overlooking the gorge. Photo by Baron Hugo van Lawick. Courtesy of the National Geographic Society.

Dr. Leakey, his wife Mary and son Philip examine tools used by prehistoric man, *Homo habilis*. Photo by Robert Sisson. Courtesy of the National Geographic Society.

Fossils, Flood, and Fire

Zinjanthropus, "East African man." The skull was somewhat like that of modern man in facial contour, although rather broad and with a strongly sloping forehead. A peculiar crest projected from the upper part of the back of the skull; however, this crest is not as prominent as that on the skull of a gorilla. The teeth are definitely human.

The most exciting feature of Leakey's work, and one whose implications do not seem to be fully realized, is that below (*sic!*) *Zinjanthropus* he found a mandible and parts of a skull and limb bones that are distinctly modern in appearance. To these he gave the name *Homo habilis.* (*Homo* is the genus name for man.) This was dated by the potassium-argon method as being 1,750,000 years old. The question is quite obvious: How could such a modern-appearing specimen lie below *Zinjanthropus?*

Other specimens closely resembling Neanderthal man were found in the higher beds. The arrangement of all of these Olduvai specimens offers a serious challenge to the whole theory of man's ascent from the apes or from ape-like animals.

A careful analysis of all the evidence regarding the African discoveries may be found in the 1966 edition of the *Encyclopedia Britannica,* if one has the patience to wade through a mass of technicalities. The conclusions that I gather from this analysis is that there is no positive evidence for the evolution of man from the apes. Note the following pertinent statement:

> Although it might appear that the primitive characters of the Australopithicinae emphasize the closeness of man to the apes . . . , these extinct creatures were already very different from apes in many fundamental hominid characters of skull, teeth, and limbs.—Art. *Man*, p. 742.

The evidence, they conclude, is that man must have evolved from a separate line than the apes. But, since these specimens are the lowest known human remains,

Dr. Leakey discovers teeth of *Zinjanthropus* at Olduvai Gorge. "The teeth were projecting from the rock face, smooth and shiny, and quite obviously humanlike," said Dr. Leakey. Photo by Des Bartlett. Courtesy of the National Geographic Society.

the evolutionists are left with no evidence as to where man did come from—no evidence of his animal ancestry.

> No fossil hominid remains have been discovered to antedate the *Australopithecus* and *Pithecanthropus;* indeed in the 1960s it was the most serious gap in the fossil record of Hominidae.—*Id.*

It is interesting to note that after more than thirty years the scientific findings have supported fully the contention of an eminent authority, when he said:

> There is a sharp, clean-cut, and very marked difference between man and the apes. Every bone in the body of a man is at once distinguishable from the corresponding bone in the body of the apes. . .
>
> Man is not an ape, and in spite of the similarity between them there is not the slightest evidence that man is descended from an ape.—Austin H. Clark, *Zoogenesis,* p. 224.

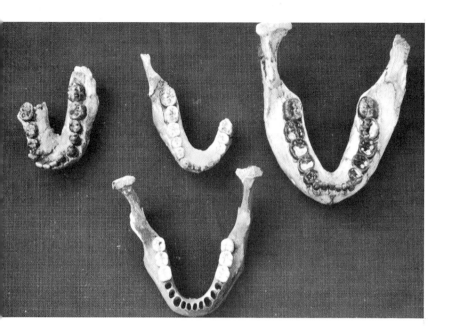

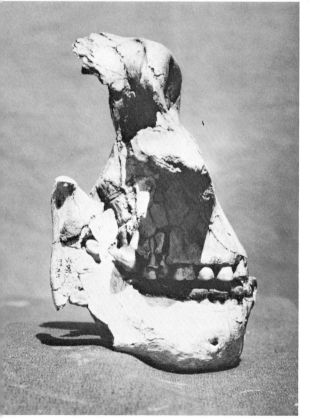

The two lower jaws (upper left and upper center) of the new species of man *(Homo habilis)* have similarities to that of modern man *(Homo sapiens,* lower center), but are quite different from the massive jaw of *Zinjanthropus* at the upper right. Photo by Hugo van Lawick. Courtesy of the National Geographic Society.

The partial skull of *Zinjanthropus* was found by Dr. Leakey and his associates at Olduvai Gorge. Dr. Leakey says: *"Zinjanthropus*...did not, as we once believed, gradually evolve in the direction of *Homo sapiens."* The flat forehead and long face are distinctly apelike, according to Dr. Leakey. Photo by Hugo van Lawick. Courtesy of the National Geographic Society.

This photo by Mary Leakey shows the back of the skull of *Homo habilis* (left) and *Zinjanthropus* (right). *Homo habilis* was discovered at Olduvai Gorge in April, 1964. Since *Homo habilis* was found much deeper in the gorge than was *Zinjanthropus*, scientists have had to revise their thinking, for it was formerly supposed that *Zinjanthropus* was an ancestor of modern man. But *Homo habilis* was a modern man as far as anatomical features were concerned, and buried deeper than *Zinjanthropus*. This places a modern-type man as the oldest known human fossil, and makes it quite clear that man did not evolve from apelike ancestors. Courtesy of the National Geographic Society.

At the left is the partial skull of *Homo habilis*, new species of man from Olduvai Gorge. At the right is the skull of a modern African. Although the skull of *Homo habilis* is smaller than that of modern man, it is certainly a modern man in all its features. Photo by Mary Leakey. Courtesy of the National Geographic Society.

Fossils, Flood, and Fire

The recent discoveries in Rhodesia are classed as Australoid, or "southern" type. This type of humanity seems to blend into the pithecanthropoid type of southeastern Asia, and it is possibly a significant point that it has been possible to trace some fossil relationships between these prehistoric skulls of southeast Asia, such as *Pithecanthropus,* and the aborigines of New Guinea and Australia.

But when we come to a consideration of the Australian and New Guinea natives, we face an important question: Are they really primitive? Possibly so, depending on what we mean by primitive. It does not necessarily mean ancestral. It does not mean that modern man has arisen from that type. No one has been able to prove, in fact, no one has even attempted to show that these people are ancestral to the "higher" types of man such as we find in Europe and America. Volumes have been written on the races of man, but when we peruse them we are struck by the tremendous amount of speculation and the dearth of substantial evidence they contain. Facts, yes, regarding the nature of the fossils, but no proofs of their relationship—only guesses. All that can be shown, according to the latest studies, is that the pithecanthropoid and australoid types are related. They constitute a distinct group, separate in many features from other races of mankind.

As evidence that the Australian natives are positively human, we might notice that they make good workmen on ranches, or "stations," as they are called in that country. They are capable of learning to work, and follow orders and operate machinery and trucks.

Schools have been established for them, and they are capable of education. They seem to have good minds. In fact, even in the wild, they have a keen knowledge of the things of nature, and are able to get along and adjust themselves to the surroundings, primitive though they may be. And they can be made into good Christains.

Yes, we may say they are "primitive." Possibly it might be better to say they have become degraded or have deteriorated, inasmuch as the word *primitive* seems to carry with it a connotation of having originated from crude ancestral stock. Without doubt they have come down from a once higher state of culture; this conclusion is just as reasonable and scientific as to assume that they arose from a beastly state.

Stone-age people exist today, right here in the United States. In the Southwest we find the abandoned dwellings of a people who once lived beneath the cliffs along the sides of the canyons. They had a stone-age civilization, it is true, but they were an intelligent people. The Pueblo Indians of New Mexico are supposed to have descended from these ancient cliff-dwellers.

Everyone knows that they are skilled people. Look at their weaving. Even though done under crude surroundings, the work is artistic, and the weavers carry the patterns in their heads and produce them without any drawings to follow.

Now is it any more correct to say that the inhabitants of New Guinea or northern Australia have an animal ancestry than to say the same for the Indians of North America? Or, turning the question around, is there any more reason for believing that some of these fossil men, because they show some "primitive" features, are of necessity intermediates between apes and man?

But, someone will say, what about the civilizations of the Middle East? Did they not come up through the different stages that are supposed to have marked the progress of mankind generally? Recent excavations have brought out some enlightening facts.

The Sumerian civilization, for example, showed no evidence of a long period of development. In the earliest tombs of the kings of Ur were gold instruments,

Fossils, Flood, and Fire

beautifully decorated. Chariots were finished in red, white, and blue mosaic, and with golden heads of lions having massive manes of lapis lazuli, and with shell on the side panels. Many vessels and ornaments of gold, silver, and copper have been found, all beautifully executed.

Excavations in Susa, the ancient capital of Persia, uncovered metal tools at the very bottom. No evidence of a pure "stone age" could be found. This was true for a number of other nearby sites. Good authorities question whether there ever was a stone age in the Middle East. A. H. Sayce, one of the highest authorities on the subject, declared that as far as archeology is concerned, it gives no evidence that the civilization of that region had any barbarous past. Apparently the sites were built by a people who were already highly civilized. This seems to be true for all of the Middle East countries.

In summarizing the whole problem of prehistoric man, it can be said that as far as long ages are concerned, only by assuming that the various flints and stone implements represent evolutionary development, could one arrive at a long chronology. The animal remains are so nearly uniform, from the lower layers to the top, that they show no evolutionary sequence whatsoever. True, there are some different animals, for instance, in the Pliocene than in the Eocene, or in the Pleistocene as compared to the Pliocene, but these differences do not prove evolutionary progression. Even though some of them may be arranged in series, that of itself does not prove they developed one from the other. The whole evolutionary interpretation is purely conjectural.

The truth is, when we put all these facts together, they portray a situation which is exactly what the Flood theory demands—a much greater variety of animal life than at present, and a mingling of types which later

Cave Man and Stone Ages

became extinct or dispersed to where they have become adapted to different climates. It may be possible that the different layers in the river terraces or in the caves may represent different stages of culture, but to assume thousands of years for the change is more than the evidence will warrant. In general the findings furnish a powerful support for the literal accuracy of the Biblical record.

CHAPTER SIXTEEN

This Changing Planet

It would be a grave mistake to assume that all geological action ceased when Noah came out of the Ark. Not only would profound changes occur in the atmosphere after the Flood, but the surface of the earth would continue to undergo shifts in many localities. Whatever cause might have been responsible for the great catastrophe (and we have not attempted to discover what it was), it would not suddenly cease to exert its effects. From all we have given so far in these studies it is evident that the violence of the action increased toward the end; obviously it would not cease immediately when the land was dry enough for man to come out on it. Perhaps it is significant that in the Genesis record Noah is said to have emerged from the Ark on high land. Without doubt the low lands and the seashores were not suitable for habitation at once.

We have a number of examples of the shifting of land levels, even in historical times. The temple at Puzzuoli, Italy, is partly submerged. The first twelve feet of the pillars above the water are smooth, but above are nine feet that have been bored by marine mollusks which live in the sands. The whole temple must have sunk below this level and then been raised

Fossils, Flood, and Fire

suddenly. At St. Augustine, Florida, stumps of cedar trees lie beneath hard beach shell-rock and are covered by the waves at high tide. In New Jersey the sea has buried entire forests in the past. On Nantucket upright stumps of trees eight feet below low tide stand with the roots still in place. Similar situations are observed in a number of other localities.

Near the coast of southern Maine at a level of 200 feet above the sea, are plain beach lines. The drift-filled valleys near Boston indicate that the sea-level must have been so low at the time of the deposition of the drift that the margin of the sea would have been out at the margin of the continental shelf, a distance of fifty miles or more. Drainage was into the "Bay of Maine," which is now entirely submerged.

Raised seashore lines are very common. Along the south Atlantic coast of the United States seven abandoned shore lines have been detected. In Spitzbergen old beaches occur as much as 100 feet above the sea. In Siberia the whole north coast has recently risen. The tundra near the coast is covered with fine clay and sand such as is now being deposited in the Arctic Ocean.

These instances mentioned are only a few of the many that could be noted from all over the world. As we consider the tremendous tectonic movement necessary to bring the present continents into position as the Flood drew near its close, it becomes apparent that the forces responsible would continue to act for considerable time.

Changes in level are also associated with the phenomena of glaciation. Continental ice masses 2000 feet or more deep would withdraw from the oceans enough water to lower the ocean level 250 feet or more. This would result in land bridges in such places as the Bering Strait or the English Channel.

This Changing Planet

The removal of the heavy load of ice from the northern part of the interior of America resulted in an uplift of the Canadian region. Many measurements indicate that this occurred in a line running from New England westward to the area west of the Great Lakes. The basin of Lake Agassiz in Manitoba was once 300 to 400 feet lower at the north end than it is now. At the Great Lakes extensive drainage changes took place. While the ice blocked the northern end of the region, the portion of the lakes comprising what is now portions of lakes Michigan, Huron, and Erie drained southward into the Mississippi River basin. As the ice melted back, the drainage changed, and took place through Lake Nipissing into the Ottawa River in central Ontario. Then as recoil lifted this region, a new drainage was established through Lake Ontario, as it is now.

Of course the geologists assume many thousands of years for these changes. But they reckon in terms of normal action. In fact, so little is known of conditions prevailing in the past that uniformitarian estimates are extremely uncertain.

Pollen analysis, one of the newest methods of determining past geological conditions, has thrown considerable light on climatic changes in the recent past. Samples of peat taken from different levels in bogs are washed and fixed by micrological techniques so that the pollen grains can be identified. We cite just one example: The lower layers in northern Europe indicate an Arctic climate, with Arctic plants, mosses, willows, and birch. As we ascend through the series of layers, pine and some hardwoods begin to come in. Eventually the profile takes on quite a modern aspect. There is clear indication of change in climate from Arctic to sub-Arctic, then to pre-Boreal, Boreal, and then Atlantic, or modern. Similar successions are indicated in other parts of the world, particularly in North America,

Fossils, Flood, and Fire

where much study has been given to the problem. The results of these studies correlate well with what we would expect at the time of glaciation and immediately afterwards.

Recent studies have shown that the advent of man into North America was correlated with the development and waning of the ice sheets. Actual camps have been found, indicating that he crossed the land bridge at the Bering Strait, or else came by way of the Aleutian Islands, and made his way down the narrow coastal strip to Mexico and Central and South America. Then, when the climate warmed up after the glacial "period," he moved northward into the center of the continent.

It doubtless is significant that we find ancient civilizations developed in a belt that was likely the first to become habitable, while many other parts of the earth were either in the grip of glaciation, or, like much of western America, under the influence of terrific volcanic activity. As history opens, we find that ancient people move farther and farther outwards towards the periphery, occupying area after area as different parts of the world became better suited for human life. There is every evidence that wave after wave of human life swept into western Europe, and there is no doubt but what the timing of these waves of humanity were influenced by the changing geological and climatic conditions.

Egypt gives a good illustration of the changes in climate that have occurred. As the country emerged into its present geographical form, it was a land of copious rainfall. The wadis were running streams, and the hills surrounding the Nile valley were covered with forest and grass. Herds of wild animals roamed everywhere. The Nile alluvium is full of shells and bones, with indications that the plain was once an extensive swamp or lake. The early inhabitants lived on the terraces along the sides of the valley, where three distinct levels

This Changing Planet

have been recognized. Highly cultured man came into the valley before the river had silted to its present level, as borings near Luxor bring up pottery.

As the ice melted away from the northern parts of Europe and North America, climatic zones began to migrate northward. There is every indication that northern types of life once existed as far south as Texas.

Musk-ox skeletons have been found as far south as Nebraska. Many large animals survived the advance of the cold, but many disappeared. Among the prominent ones at this time were the mastodon and the mammoth.

As the Pleistocene glaciation passed, the earth rapidly emerged into its present state. But the vast herds of pre-Pleistocene animals had largely disappeared, except in central Africa, where remnants of them still exist.

Not only has there been climatic migration, but also a "progressive dessication," that is, a gradual drying up. Interior basins once filled with water, such as the basins of Lake Lahontan and Lake Bonneville in Nevada and Utah, the Caspian basin, and parts of the Sahara, either became dried up or the waters shrank in size. Regions of the Near and Middle East that are now almost entirely desert were once the site of prosperous civilizations. Portions of western America apparently once supported agriculture to a much better degree than is possible today without modern methods of irrigation. Furthermore, there is clear evidence that the average annual temperature of certain regions is slowly, but surely, rising.

Much more detail might be given along these lines, but perhaps enough has been given to make the point clear that the catastrophic interpretation of geology has much that is worthy of attention. Again we should be reminded that *uniformitarianism is unproved and unprovable*. To interpret the past in the light of this hypothesis is to follow assumptions for which no scientific

Fossils, Flood, and Fire

support can be given. On the other hand, evidence everywhere indicates that the earth has undergone terrific violence, which can be explained only in terms of the Great Catastrophe.

This does not mean that there are no problems to confuse us. There are many, and some of them are very puzzling. On the other hand, so are there problems in the uniformitarian hypothesis. The conclusion a person will reach will depend largely on his basic philosophy. Having established that, he will interpret details in its light, and lay aside unsolved problems for further study. But he will not allow the problems to divert his attention from the main lines of evidence he has accumulated to explain the phenomena of geology according to his basic philosophy.

At this point it might be well to answer a question that may linger in the mind of the reader as he comes to the end of this presentation of the theory of Flood geology. It is this: What recognition do you expect for this theory from the scientific world? In answer we may say that we are not concerned about that. The confirmed believer in uniformitarianism will not see any sense in our interpretation. One who is not clear in regard to the questions involved may be able to give serious consideration, and possibly may be led to realize the scientific validity of the Flood theory. But whatever the reaction, we are not going to be disturbed over the question of recognition. What we have tried to do is to present geology as it appears to one who is firmly convinced that the Genesis account is valid. This is not to say that we have been correct in every interpretation. We have tried to check our technicalities carefully, but even with the best that can be done, there may be some errors in the presentation. If such are found, we shall appreciate having them pointed out. But if there is a difference of opinion merely on interpretation, then we shall demand that the catastrophic interpretation be

This Changing Planet

given its rightful place and weighed against the uniformitarian view. We are sure that when the evidence is all in, as the jury would say, there will be much to be said in favor of the Flood theory of geology. That is all we ask or hope for. This study will be a base from which further details can be developed, until eventually we shall have a real science of "Flood Geology."

General Index

Figures in italics refer to illustrations

A

Absaroka Range 155
Aborigines, Australian 222
Adaptation *vs* evolution 167
Agassiz, Louis
  Glacial theory and the Flood 18, 169
Agassiz, Lake 187, 188, 229
Alaska 155, *158*, *159*, 201
Alexandrian schools 14
Algonkian 67-9, 71-4, 77, 136-7
Alps 155
Amethyst Mountain, Wyoming *118*, *121*
Ammonites 132, *133*
Amphibians 100-101
Ape ancestry of man 219
Appalachians 74, 91, 93, 96, 104, 112, 132, 146
Aquinas, Thomas
  Dualism between science and religion 16
Arabic culture
  And creation 16
  And European learning 16
Archaean rocks 67-9, 71, 74
Arches National Monument *115*, *122*
Aristotle
  Evolutionary views 13, 14
  Influence on Arabic learning 16
  Influence on Christian philosophy 16
Armored fishes 83
Arthrodire *78*
Athabaska Glacier *192*, *193*
Atlantic coastal plain 70, 104
Atlantic Ocean "mountains" 140
Augustine, creationist views 15
Australian "blacks" 222-3
Australoid man 222
Australopithecus 214, 219
Averroes 16

B

Bacon, Francis
  Dualism between science and religion 17
Bancroft, H. H. 45
Banff, Alberta 137
Basalt *156*
Basement complex 65ff, 70
  See also Algonkian, Archaean, and pre-Cambrian
Basins in Rocky Mountain region 132
Batholiths 134-5, 138
Belemnites *133*
Beltian system 69
Bible
  Relation to science 18-19
  Inspiration 18
Bible-Science Association 44
Big Horn Mountains 150
Black shales 87
Blastoids *78*
Blue Ridge Mountains 70, 104
Bog theory of coal formation 33
Bonneville, Lake 231
Brachiopods *78*, 79
Breccia *157*
Bridges, land 161
Brontosaurus 130
Bryozoa 81
Buffon and the "day-age theory" 18, 21, 22

C

Calamites *98*
Calcott, Alexander 38
Cambrian 25, 35, 55, 69, 71, 77, 79, 81, 83, 85, 86-7, 92, 93
Camels, fossil *165*, 168
Canadian Shield See Shields
Canadian Rockies, rocks of 72-3
Canyonlands National Park, Utah *117*, *148*
Capitol Reef, Utah *119*

General Index

Carboniferous 26, 92
Catholic Counter Reformation 17
Cascades (Cascade Mountains) 155, 172
Cascadia, ancient land of 85
Cave men 205ff
Cenozoic 26, 143ff
Cephalopods 78, 133
Challenger Expedition 30
Chief Mountain, Montana 136
Chinle formation 116, 120-1, 125
Christianity, influence of Greek culture on 14-15
Chronology, long vs short 7, 42
Cirques 191
Classification of rocks, origin of methods 25
Climates in past 161, 172, 180, 186, 208, 225, 229-31
Coal beds, formation of 33-5, 91ff, 114, 123, 125, 154
Colorado Plateau 32, 93
Colorado River 148, 149
Columbia Plateau 154
Corals 78, 79
Cordaites 99
Correlation, methods of 26-7 57-58, 86
Creation
   And revelation 13
   Augustine's views 15
   Supernatural 12
   vs evolution 11
Creationism
   And the Reformation 17
   Attitude of science on 11
   "Golden Age" of 17
   Modern view on 18
   Oldest views in existence 12
   Orthodox belief of Hebrews 12
   "The New Creationism" 18
Creation Research Society 44
Creation Week, no life before 42
Cretaceous rocks 25, 111, 114, 123, 127, 132, 133, 134, 136-9, 143, 150, 160
Crinoids 78, 81, 83
Cro-Magnon man 209, 212, 213
Cross-bedding 31, 32, 122
Crust of earth 65, 88
Crossopterygian fishes 78
Cuvier's catastrophism 18, 32, 33
Cyclic sedimentation 32
Cyclothems 92, 112

D

Darwin 40
Dating, methods of 158
da Vinci, Leonardo
   Views on fossils 37
Delicate Arch, Utah 115
Deluge legends 45ff
Devonian rocks 25, 77, 81, 82-3, 85, 87, 91-3, 111, 154
Diastrophism 106
Diluvialism 8, 9, 29, 61
   See also Flood, Flood Theory of Geology
Dinosaurs 11, 35, 120, 128-30, 131, 160
Diplodocus 130, 131
Dragonfly, fossil 98
Drumlins, glacial 173, 175

E

Earth, structure of 65-6
Echinoderms 35, 133
Ecological zonation theory 91, 102
   Similair to geological correlation 57-8
   Suggested by Price's zoological provinces 53
Ecological zones See Zonation
Elephants, fossil series 168
Emergences 95
Eocene 144, 145, 224
Eohippus 162, 166
Epic of Gilgamesh 46
Equisetaceae 100
Equus 162, 163, 165
Erosion of Colorado Plateau 125
Erratic boulders 173, 176, 177
Eskers 172
Evolution
   Origin of theory of 13
Evolution, Twilight of, by Henry M. Morris 43

F

Ferns 100, 127
   Seed ferns 100
Fish
   Armored 83
   Crossopterygian 78
   Fossil 35
   In Green River shales 147
   Pennsylvanian, fresh-water 100

Fossils, Flood, and Fire

Flood
    Attitude of scientists toward 12
    Changes after 227ff
    Close of, when? 144, 155
    vs geology 7
    Influence of Hutton's theory on belief in 23
    Legends of 45ff
    Reformation, attitude toward 17
    Universal catastrophe 42
Flood theory of geology 18, 37ff, 43, 61, 64, 95, 233
    Resurgence of 44, 61
    Revived by George McCready Price 40-41
    Why rejected 39
    See also Geology
Forests, fossil See Fossil trees
Fossils
    And "age" of rocks 86
    Burial of 43
    Early views regarding 21, 37
    Index 86
    Order of (not dependent on "age" 55
    Paleozoic 78
    Price challenged sequence of 53
    Remains of ancient life 21
    Typical Paleozoic 78
Fossil fish 147
Fossil trees
    Arizona *119*
    South Joggins, N. S. *62, 63*
    Through many layers or coal 96
    See also Trees, fossil
    Yellowstone Park *118, 121*
Fusulinids 102, *103*

G

Gastropods *78, 133*
Genesis account of Flood
    Account valid 232
    And science 7, 11
    And revelation 12
    Influence of rise of geology 18
    Older than Greek theories 12
Genesis Flood, The, by Whitcomb and Morris 43
Geologic Systems 52
    Survey of 55ff
Geology
    And the Flood 7
    Influence on interpretation of Genesis 18
    Influence of modernism on geological interpretation 40
    Modern Flood theory of See Flood theory of geology
GeoScience Research Institute 44
Geosynclines *66*, 77, 84, 85, 91, 136
Gilgamesh, Epic of 45, 46
Glaciation 169ff, *171, 173, 175-196,* 230
    Agassiz, views and in relation to the Flood 18
    Map of *171*
    Permian 106ff
    Supposed Permian 107ff
    Thickness of ice 180, 182
    Temperature of ice 182
    Time of 178, 201
Glacier National Park 70, 136
God's, personification of natural forces 12
"Golden Age" of creationism 17
Grand Canyon strata *56*
Great Lakes and glaciation 187, 229
Greek culture and Christianity 14
Green River formation *148, 149*
Green River basin 147
Grooves, glacial *180, 183, 185*
Guadalupe Mountains 106

H

Hall, James 26, 105
Heidelberg man 210
Himalayas 155
Historical Survey 9ff
Homo habilis 217, 218, 220-21
Horses, fossil *162-165,* 166, 167
Hutchinson, John 38
Hutton, James
    Theory of the Earth 22

I

Ice sheet See Glaciation
Ichthyosaurs 130, *131*
Index fossils 86
Insects
    Fossil *101*
    In Jurassic 127
    In Pennsylvanian 100, *101*
Ionian philosophers 12, 13

J

Jehovah different from other gods 12
Jellyfishes 79
Joggins, N. S., coal beds in 34

General Index

John Day beds, Oregon *150, 152,* 153
Juarez 17
Jurassic rocks 25, 111, 114, 122, 127, 128, 139

K

Kames 172, *175*, 186
Katmai, vulcanism in *158, 159*
Kettles, glacial *173*
Kishkawalsh Glacier *195*

L

Lahonton, Lake 231
Lava flows
  Oregon and Yellowstone *156*
Leakey, L. S. B. 214, *216*, 217
Lehman, J. G., classification of rocks 21
Lepidodendron, scale tree 97, *98*, 99
Life zones, ancient compared to modern 36, 51ff, 58
Llanoria, ancient land of 106
Lycopods *98*
Lyell, Charles
  Opposed by some 51
  Principles of Geology, first textbook 24
  Influence on geological theory 24

M

Magmas 135
Mammals, first appear in Cretaceous 134
  "Evolutionary series" of 161
  Fossil 36
Mammoths 178, 188ff, *196*-7, *198, 199, 200*, 201
Marl deposits 138, 148
Matter, eternality of 12
Merychippus *162*, 166
Mesohippus *162, 164*, 166
Mesozoic rocks 26, 55, 111ff, 127, 134, 139, 141, 144
Metamorphism 74
Middle East civilization 223
Midwest of U.S.A., ancient geography of 87
Migration of animals 161
Miocene 144
Miohippus *162*, 166
Mississippian rocks 26, 55, 83, 86, 91, 92-3

Missouri River and glaciation 186, 187, *188*
Modernism and geology 40
Moenkopi formation 116, 125
Mollusks 35, 132
Monument Valley, Utah 31
Moraines, glacial 170, *177, 179,* 186, *192, 193, 194, 195, 196,* 206
Morris, Henry M. 43
Morrison formation 122-3, 128
Mososaurs 130, *131*
Mount Stephen, British Columbia 80
Murchison, Roderick 25, 83, 105
Myths and legends of the Flood 45ff

N

Naturalistic philosophy 13
Nature worship 13
Neanderthal man *209*, 210-12, *211*
Nelson, Byron C., *Deluge Story in Stone* 38, 48
New Creationism 18
New Diluviaslism, interpretations in 41
New Geology, by George McCready Price 7
New Guinea natives 222-3
Nomenclature, origin of geological 25
Northwest, conditions in 88

O

Ohio River and glaciation 187
Olduvai Gorge, Africa 214, *215, 216, 217*, 218, *219*
Oligocene 144
"Onion-coat theory" 22
Ordovician rocks 25, 71, 77, 81, 83, 86-7, 89, 91, 92
Orohippus 166
Ooze on ocean floor 30
Overthrusts 104, 136ff
Ovid 47
Oysters 132

P

Paleocene 144, 145, 160
Paleolithic "age" 205ff
Paleozoic rocks 25, 55, 56, 74ff, 77, 88, 116
Parahippus 166
Paris Basin, sedimentation in 33

237

Fossils, Flood, and Fire

Pelecypods 78, 133
Peneplains 68
Pennsylvanian rocks 26, 34, 55, 84, 86, 91-2, 93, 94, 95, 100-2, 127, 139
Permian rocks 25, 55, 82, 84, 105ff
Periods, geological *vs* Flood stages 84
Petrified Forest, Arizona *118*, 120
Pithecanthropus 208, *209*, *211*, 219, 222
Plants, fossil
  First appearance in Devonian 92
  In Cretaceous 127
Plato and Christian theology 13, 16
Playfair, John, and Hutton's theory 23
Pleistocene 26, 144, 169ff, 224, 231
Plesiosaurs 130, *131*
Pliocene 144, 172, 224
Pliohippus *162*, 166
Polish, glacial 170, *180*, *181*, *183*, *184*, *185*
Pollen analysis 229
Pre-Cambrian 68, 70, 79, 85, 112-14
  See also Algonkian, Archaean, Basement Complex
Prehistoric man 206ff, *207*, *209*, 224
Price, George McCready, 7, 39
  Attacks evolutionay geology 53
  Revision of views 42
  Revives "Flood Geology" 40, 41
Properties of matter 14
Protozoans, fossil 102
Pterodactyls 130, *131*

Q

Quaternary 143

R

Rainbow Bridge, Utah *113*
Red beds 111ff, *113*, *114*, *117*, *118*, 134
Reefs 81-2, 106, 132
  Oil produced in 81
Reformation
  And creation 17
  And relation of science to religion 17
Reptiles 35, 55, 100, 129-30, *131*
Revelation
  And science 9
  And Genesis record 12
Rift Valley of Africa 140

Rocks
  Early classification of 21
  British classification 25
  New York system 26
Rocky Mountains 132, 154-5, 174

S

Salts, deposits of 110
Sandstone, cross-bedded 31
Sandstone in coal beds 34
San Juan River, Utah *149*
Scale-trees (Lepidodendron) 97
Schists 67-8
Science
  And the Bible 18
  And revelation 9
Seas, ancient 59, 128
Sedgwick, Adam
  Retiring address opposes uniformity 51, 53
  Studies on rocks of England and Wales 25, 83, 105
Seed ferns 100
Shales, black 87
Sheep Mountain, Wyoming 71
Shields, *66*, 67
  Canadian 68, 69, 74, 84, 85, 94, 102
  Cordilleran 85
Shinarump formation 116, 120, 124
Siberia, fossils in 201
Sigillaria 99
Silurian rocks 25, 55, 74, 77, 81, 83-4, 92
Sinanthropus *209*
Smith, William, fossils and strata 23
Socrates 13
Species, validity of fossil 167
Spencer, Herbert, attacked Werner's "onion-coat theory" 51, 53
Sponges 79
Starfishes 79
Stegosaurus 130, *131*
Stenson, Niels 21
Stigmaria 99
"Stone ages" 205ff, 223
Striae, glacial 170, *183*, *184*, *185*
Submergences 95

T

Tectonic movements 134-5, 138, 140

238

General Index

Tertiary rocks 26, 57, 143, *151*, *152*, 155, 160, 172, 174
  Animals and plants of 145
  Classification of 144
  Climate of 146
  Of Guld Coast 146
  Of Rockies 147
Texas, sedimentation on coast of 33
Thrusts See Overthrusts
Till 172
Tillites 106ff
Titicaca, Lake, rocks of 75
Trees, fossil *62*, 96-7, *97*, *119*, *121*
  See also Fossil trees
Triassic rocks 25, 111, 112, 114, 116, 122, 139
Triceratops *131*
Trilobites 35, *78*, 79, *80*, 81
  Olonellus fauna 79
Twilight of Evolution, by Henry M. Morris 43
Tyrannosaurus *131*

U

Uniformitarianism
  And diluvialism 30, 36
  False theory 42
  Fundamental to modern geological theory 63, 106
  Hutton and Lyell 23, 24
  Problems of 30
  Why accepted 39-40
  Unprovable 232
Uniformity
  No evidence on Colorado Plateau 125
  Problems regarding ancient reptiles 130

V

Varves 202
Violent burial of plants and animals 35ff
Vishnu schist 68
Volcanic activity See Vulcanism
Vulcanism 149, *157*, *158*, *159*, 174

W

Water, evidence of violent action of 32-3
Wave action
  Effect on life zones 59, 96
  On Colorado Plateau 124
Werner, A. G., "onion-coat theory" 22
White Mountains, N. H. 89
Whitcomb, John C. 43
Williams, John 38
Wind River Mountains, Wyoming, stratigraphy of 54, 150
Wisconsin ice sheet 186
Woodward, John, views on fossils 37
Worms 79

Y

Yosemite glaciation 170

Z

Zinjanthropus 218, *219*, 220-1
Zion National Park, Utah 31
Zonation in ancient world 36, 42, 51ff, 58, 127, 129, 160
  See Ecological Zonation Theory
Zoological provinces or zones
  See Ecological Zonation Theory

239

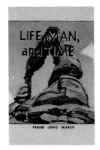

Life, Man, and Time

by

Frank Lewis Marsh, Ph.D.
Professor of Biology
Andrews University

This revised edition of a former book by the same title deals with such topics as Age of the Earth, Cave Men, Human Fossils, Radioactive Dating Methods, Carbon 14, and many related subjects. The frank discussions are written for the layman, yet are scientifically accurate. Dr. Marsh has spent a lifetime as a biology teacher in two major colleges, and has had a great deal of experience dealing with the many phases of the controversy between creationism and evolutionism.

The 14 chapters in this book are—

1. Fundamentals
2. Historical Theories of Origins
3. The Genesis Account of Creation
4. Age of the Earth and Its Inhabitants
5. Though Formed of Dust, A Son of God
6. A Controversy in Nature
7. A Threefold Curse
8. After the Flood
9. And the Earth Brought Forth Plants
10. And the Earth Brought Forth Living Creatures
11. Variation Among Organisms
12. Fixity Among Living Things
13. Ancient Men
14. In Conclusion

This book is unique in that it shows how it is possible to believe in the inspired record of the Holy Bible and yet be entirely scientific in attitude and study. Dr. Marsh explains the world and life upon it without recourse to evolution. Evolutionists as well as creationists will want to read **Life, Man, and Time.**

Published in 1967, illustrated with photos and drawings, $4.95 postpaid.

Order from OUTDOOR PICTURES, Box 1326, Escondido, Calif. 92025